PURNELL'S ENCYCLOPEDIA OF
INVENTIONS

PURNELL'S ENCYCLOPEDIA OF
INVENTIONS

PURNELL

EDITORIAL

Editor: Peter Presence

Contributors:
Michael Bucknell
Arthur Butterfield
John Clark
A. W. Coysh
Dorothy Dixon
Bernard Moore
Peter Muccini
Theo Rowland-Entwistle
Jonathan Rutland

Designed by Eric Rose and John Perrett
for the Queensbury Design Group Limited

Published 1976 by Purnell Books
Berkshire House, Queen Street, Maidenhead, Berks.
Produced for Purnell Books by
Intercontinental Book Productions

Copyright © 1976
Intercontinental Book Productions

Printed by Purnell & Sons Ltd., Paulton, Bristol

SBN 361 03495 4

CONTENTS

'Puffing Billy', William Hedley's locomotive of 1813 which proved that metal wheels could grip on metal rails.

INTRODUCTION

Anyone attempting to compile a complete inventory of all the world's inventions would have a hopeless task. The patent offices of the Earth's major countries are crammed to bursting with the ideas of eager scientists and others, many of which have drifted there only to perish in the archives, like prehistoric flies trapped in amber. And for every idea that has been patented, there are probably half a dozen that have not; brainwaves every one, and some of them of world-shattering importance.

The wheel, for instance: the great mechanical force that has no counterpart in Nature. Neolithic Men 5000 years ago used rounded stones on which to trundle the huge blocks of stone with which they built their temples: the stone rollers still lie deep in the soil under the megaliths they helped to transport. It seems simple device, obvious to anyone: yet the great civilizations of South and Central America, the Aztecs, Incas and Maya, did not have the wheel, though they had invented their own form of writing and numeration.

What sparks off an invention? The causes vary. Some people are inventors by nature: the American Thomas Alva Edison, for example, is credited with more than 1100 inventions patented — and heaven knows how many more which ended up in the wastebin.

Most people invent what happens to be needed. A man who has to find a way to do a job will invent a tool or a method to do it. In this way James Starley, faced with the problem of how to have the two rear wheels of his tricycle moving at different speeds, while maintaining a drive on both, devised the differential gear. Sir Humphry Davy, asked to help miners by inventing a lamp that would not cause explosions, did so in the incredibly short time of five months — and would not take a penny for his pains.

On the other hand, many people have a dream, and spend their lives trying to bring it to life. Scotland's John Logie Baird, sure that television was possible, toiled for years alone and with minimal financial help to make it so. One of the men responsible for the typewriter, Christopher Lathom Scholes, even spent months on his deathbed making more and more inventions for his beloved machine. During World War Two, Ronald Hamilton, a Briton who just could not stop inventing, spent his days in a disused wing of London's Grosvenor Hotel producing, among other things, a floating roadway, which saw service in the Allied invasion of Norway in 1944.

This book gives a comprehensive view of the world's most important inventions in the fields of transport, communications, industry, warfare — and also in the home and the office, places where people spend so many hours which are made bearable thanks to other people's brainwaves.

The story of the inventors is one showing great changes not only in the world around them, but also in the way inventors work. In earlier days, a man would sit down with a piece of paper and a pencil, or tinker away at his workbench — and lo! there was another invention. Today, the every-increasing complexities of life mean that hardly any inventions are one-man efforts; generally they are the work of teams, sometimes very large teams indeed, of scientists of many abilities.

The Beginnings

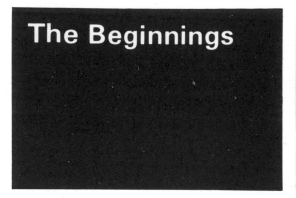

Charcoal, one of Man's first 'prepared' fuels, was burned in open heaps like this until comparatively recent times.

It is hard today to imagine the state of Early Man on Earth. In the beginning he had no home, no tools or weapons, no clothes. He had not learned to grow food for himself, and depended solely on his skill as a hunter, and on gathering such edible fruits and nuts as he could find. In finding food, Man was little different from the animals around him.

But there was one important difference. He was endowed with a more advanced brain, a brain which made him inventive and resourceful and was to bring him, through the ages, to astonishing levels of achievement far beyond comparison with that of any other animal life.

Early Man was the species known as *Homo erectus* – 'Upright Man'. What Early Man first 'invented' will never be known. Probably that first invention was a primitive tool made from a split stone which served a number of purposes, especially skinning and cutting up his food. Many such tools, half a million years old and more, have been discovered by archaeologists.

From these simple tools Man developed the hand axe, and made a simple kind of knife by flaking stone to a sharp edge. Much later he found that by attaching the hand axe to a rough shaft of wood he acquired greater leverage and had made not only a more efficient tool but also a relatively effective weapon – the axe as we know it today.

When Man first discovered how to make fire is not known. It was probably by accident. Traces of man-made fire 500,000 years old have been found in China. Fire was one way of keeping warm, but with his more developed hand-tools Man was able to use the skins of the animals he killed to make clothes. And although most people lived in caves, they were quick to teach themselves to make primitive huts with skins and the branches of trees.

Homo erectus was succeeded by Neanderthal Man about 70,000 years ago. About 40,000 years later *Homo sapiens* – Modern Man – had become the dominant species. With the emergence of Modern Man inventions became infinitely more sophisticated.

By 10,000 BC Man had begun to organize and ensure his supplies of food. He had discovered farming, the growing of crops, and from that discovery developed the invention of agricultural tools. At first these were primitive and operated by hand, but soon he made more sophisticated instruments such as the ox-drawn plough, for he had learned both how to use metal and how to domesticate animals.

Exactly when and where Man made one of his most important inventions – the wheel – is not known. It may have been in Sumeria, in the Middle East, for the 5000-year-old Standard of Ur, an inscribed stone which depicts many Sumerian customs, shows a four-wheeled chariot. The wheels are solid semi-circles of wood pegged round the axles. Spoked wheels were apparently not invented for another 1000 years.

The development of farming led to more settled

Left: The saw was one of the earliest tools of civilized Man. This one was shown on a tomb built about 1480 BC.
Below left: Fire drills, among the earliest means of making fire, have persisted into modern times, as with Australian Aborigines.
Below: A primitive reed canoe, an example of how Early Man used to travel on water.

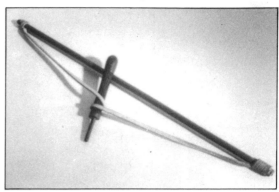

Above: Spears and spear throwers — hand-carved wooden implements — were among the first weapons used for hunting, and also for warfare.
Above right: The bow drill was a more sophisticated form of fire-maker.
Right: Simple yarn-making, as done by this Australian Aborigine, is a skill which has been handed down since prehistoric times.

communities which, in turn, led to permanent forms of shelter. More solid houses took the place of the huts and tents which had sufficed the nomads. Man discovered how to make bricks, probably observing that the hot Sun baked mud hard, and it was a short step to shaping the mud so that it could be easily handled.

The remains of Mohenjo-Daro, a city of the Indus Valley civilization of Pakistan, which flourished some 4500 years ago, show that it was a city built on a gridiron plan, with wide streets whose buildings were made of uniform kiln-dried bricks, superior to those used earlier in Sumeria.

With more permanent shelter and a more assured supply of food, Man's thoughts turned to creature comforts such as clothes, cooking pots and other vessels, and he mastered the handling of metal even to making personal ornaments such as rings, bangles and necklaces.

Pottery, the shaping of clay and the baking of it, was probably one of Man's earliest discoveries. The clay was first moulded simply by the potter's hand, but the invention of the potter's wheel led to vessels of greater symmetry and beauty. Such pottery was being made by the Sumerians 6000 years ago.

Spinning and weaving date from Neolithic – New Stone Age – times and are known to have been used 8000 years ago in western Asia. Textile manufacture was woman's work. Thread was spun from flax or wool on a spindle twirled in the hand. It was then woven into cloth on primitive looms. The ancient Egyptians of 4000 years ago wrapped their dead in fine linen woven from flax grown along the banks of the River Nile.

At a very early stage Man also used his remarkable brain to conceive inventions not obviously connected with his physical life. He devised the alphabet and learned to write and read. He invented devices for measuring the passage of time and worked out systems of counting and calculation which we know as mathematics.

Writing was developed by the Sumerians, as we know from a 5500-year-old tablet found in Mesopotamia. The inscriptions were incised on stones or soft clay which was then baked. Writing at first took the form of crude picture-writing which later developed into what became known as cuneiform writing, from the wedge-shaped symbols used. Other civilizations, such as those of the Indus Valley and the Chinese, devised forms of writing at least 3500 years ago, and the Egyptians had their system of picture-writing which we know as hieroglyphics.

These ancient systems of writing involved the use of hundreds of symbols, but gradually more economical systems were devised, more like the alphabets of today. By about 1000 BC the Phoenicians had an alphabet of 22 signs, which the Greeks modified to 24 letters. It is the basis of our 26-letter alphabet.

Paper was 'invented' by wasps, as you can see if you examine a wasps' nest. It was the Chinese who first copied the insects successfully, around 2000 years ago. But much earlier the Egyptians broke away from baked clay tablets by using the reed papyrus, whose stalks were shaved into thin slices and pressed into sheets not unlike paper.

The measurement of time began with observation of the apparent motion of the Sun, a natural clock, and the ancient Egyptians devised a primitive sundial using an upright pole and a crossbar. Later the Egyptians, Greeks and Romans invented water clocks which could be used indoors and did not rely on the Sun.

The seven-day week and the four-week month were devised by the Babylonians, and a 365-day year was recognized by the ancient Egyptians.

With the development of more settled communities, man's versatile brain turned towards methods of counting and calculation. Some mental record was needed for the control of herds and trade. From finger-counting, almost certainly the first form of arithmetic, were developed more sophisticated mathematical systems leading to the abacus of ancient China, Greece and Rome.

These early systems of numeration took account only of actual numbers – there was no way of indicating nothing. An outstanding invention in mathematics was the symbol for zero, devised in India 1500 years ago. This symbol did not reach Europe until the Middle Ages.

We take most of these inventions for granted today. But considered in the context of their time, many of them were no less revolutionary than the invention of the steam engine, radio, spaceships, and the countless other examples of Man's inexhaustible ingenuity, which are described in the following pages.

Communication

Creatures are made aware of their environment through the senses. Taste, smell, and touch give warning of danger. Sight and hearing help animals to communicate with fellow-members of their species. Birds have their courtship displays, their songs and alarm calls. Most mammals have mating calls and cry or squeal in the presence of danger.

Civilized Man may derive great pleasure from the senses of taste, smell and touch, but communication depends largely on sight and hearing. We can tell whether someone is friendly or antagonistic by the look on his face and the sound of his voice.

In the 1600s philosophers became interested in the way the senses of sight and hearing operate. It was Galileo Galilei who discovered that the pitch of a note depends on the number of vibrations in a unit of time, a measurement now known as 'frequency'. He realized that ordinary sounds, as well as speech, create waves of compression in the air and when the vibrations reach an object they cause this, in turn, to vibrate. In the human ear the vibrations are concentrated on the ear-drum which transmits vibrations to the membranes that make hearing possible.

Normally the human ear can react to a range of some seven octaves of tones. But there are sound vibrations, such as the cry of a bat, which are outside the range of the human ear. A Wittenburg philosopher, Ernst Florens Friedrich Chladni, was the first person to study sound vibrations in detail, in the early 1800s, and he is usually regarded as the father of the science of acoustics. Most inventions concerning sound have been based on his discoveries.

When scientists realized that sound consists of vibrations, they began to speculate on the nature of light. Sound can travel round corners; light, however, seemed to travel only in straight lines. Robert Hooke was the first man to put forward a wave theory of light. He published his views in 1665. A more complete theory was advanced in 1678 by Christiaan Huygens, a Dutch mathematician, though he was unable to explain the phenomenon of refraction — bending of light.

In 1801 a second wave or undulatory theory of light was advanced by Thomas Young, and, after discussion among scientists over a period of nearly 50 years, it was generally accepted.

Human experience is built up largely from what is seen and what is heard. Most inventions in the field of communication extend the range of these impressions. Today, communication is possible over vast distances; we have even been able to see and hear men landing on the Moon. Scientific invention has given us the telephone, radio, television and radar, extending the range of our sight and hearing — a process that has taken centuries.

Primitive Man realized the need to communicate with other members of his tribe or group at times when he found himself a considerable distance away from them and not in a direct line of vision. Sometimes he sent a signal by lighting fires and putting up smoke trails. Alternatively, he made some distinctive noise to send a message — the sound of a flute or horn, or the beat of the 'jungle drums'.

Such methods were important if speed of communication was essential, but it was found that a message could be sent without the sender being seen

Above, top: Signalling with flaming torches was an early form of communication.
Above: A flaring beacon on a hilltop served to give warning of danger.
Top right: Cave painting, one of the oldest forms of communication known — the 20,000-year-old so-called 'Circus horse' from the cave of Le Portel, in southern France.
Below, right: Drums like this have been used for signalling in parts of Africa for centuries.

Opposite: Hieroglyphics — pictorial symbols — formed the earliest kind of written communication. This inscription comes from a tomb at Thebes, in Egypt.

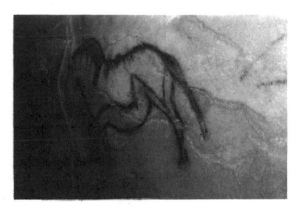

or heard. Some form of symbol was involved, which might take the form of a knot in a cord or a notch cut in a stick. Such symbols could be carried from one person to another by an intermediary. They could also be used to keep records. All manner of objects have been used for this purpose – sticks, bones, pebbles and beads of all sizes and colours.

The use of a tool to make a crude design on a weapon or a piece of pottery eventually led to more sophisticated engravings or paintings on rock surfaces or on cave walls, usually of the animals which were hunted for food. Some of these pictures were no doubt meant to convey messages and they are often referred to as pictographs. A series of pictures was sometimes intended to tell a story. As this method of communication developed, the pictures gradually became standardized, and were widely accepted and understood as symbols.

At first, the symbols represented physical objects in the environment, but a new need slowly began to emerge. As people advanced, they began to take an interest in ideas and the supernatural. The symbols had then to be made to represent thoughts and feelings, as well as natural objects. New ways were found of expressing these. A symbol showing water flowing from a jar, for example, would convey more than the physical objects drawn. It would give an impression of coolness, especially in a desert land. Footprints pointing towards a dwelling might be meant to represent an invitation.

The use of symbols in a logical sequence led to the earliest forms of writing. Clay provided one of the most suitable surfaces on which to draw the symbols. Wedge-shaped impressions were made in moist clay with a stylus of reed or wood, and it was then allowed to dry in the heat of the Sun. At first, the writing was made on small tablets of clay which could be held in the hand. Later, larger tablets were used which were laid flat on a bench or table. Such tablets could be preserved as records, and many have survived to this day. The writing is usually known as cuneiform, since the impressions of the stylus were wedge-shaped or cone-shaped.

The earliest cuneiform system of writing was used by the Sumerians, who lived in Mesopotamia about 3000 BC on a stretch of land between the Tigris and Euphrates rivers. They developed writing to a fine art, and a class of professional cuneiform writers, known as scribes, emerged. Many of the symbols proved to be unwieldy and the scribes slowly simplified the system, reducing the number of symbols and representing sounds instead of objects. Thus the first phonetic system of writing developed.

The use of cuneiform writing slowly spread throughout the Middle East; tablets have been found with cuneiform symbols of the Babylonian, Assyrian and Hittite languages. Similar writing developed independently in China and in Egypt. The early picture-language of hieroglyphics used by the priests in Egypt also developed into a system of writing based on sounds.

The early writing systems of the Sumerians and Egyptians were adapted for their own purposes by several Mediterranean peoples. The Phoenicians were

particularly methodical, and invented a script having 22 signs. The first was *aleph* (the Phoenician for 'ox'), the second was *beth* (the Phoenician for 'house'), and so on. Not surprisingly, since the Phoenicians were great traders, their system was carried to other countries, where it was developed to suit local needs. The Greeks added vowel sounds and started their symbols with *alpha* and *beta* – hence the origin of our word 'alphabet'.

As more and more written work appeared, clay tablets began to prove too cumbersome. The first paper as we know it today was probably produced in China in AD 105, when plants were beaten into fibre and then reconstituted to form a sheet. The Chinese also used hemp, cotton and the bark of trees. For ink they used cochineal insects and tree sap, but by AD 300, they had produced a black ink using lamp-black and soluble gums.

In Egypt it was found possible to record hieroglyphics on the pith from a plant known as

Left: For centuries, books were produced, one copy at a time, by scribes, mostly in monasteries. *Above:* Papyrus, one of the earliest forms of writing material, was gathered from plants growing by the banks of the Nile, as shown in this ancient Egyptian picture.

Clay tablets were used in Assyria 5000 years ago. This one is inscribed with incantations to a fire-god.

Right: Watermarks used by early German papermakers to identify their wares.

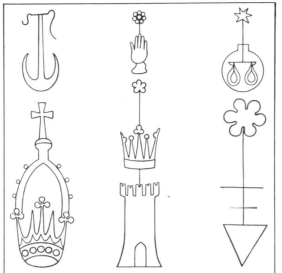

Far right: Paper, first produced 1800 years ago in China, was made by hand until the 1700s. Here is papermaking as it was carried on in 1568.

papyrus which grew freely by the Nile – hence our word 'paper'. The pith was flattened into sheets. The production of papyrus became a considerable industry in Alexandria, where Alexander the Great had founded a library. The writing was done with a thin reed or calamus. One end of the reed was bruised so that the fibres made a fine brush; later the reed was cut to form a point for writing.

In the 100s BC a library was formed at Pergamum in Asia Minor, but paper could not be obtained from Egypt because the Egyptian ruler, Ptolemy, feared competition with his library at Alexandria. This led to the invention of a new writing surface – parchment, or *charta Pergamena*. It consisted of the specially prepared skin of a calf, sheep, goat or pig.

First, the surface hair was removed. The skin was then left for some time in a lime pit and afterwards stretched on a frame. The surface was then scraped on both sides and treated with chalk. The finest parchments were known as vellum, from the Old French word for calf.

Papyrus and parchment could be preserved in scrolls. These consisted of many strips of parchment pasted together to form a long sheet, which was then attached at each end to a wooden rod so that the sheet could be rolled up from either end. scrolls up to 100

feet (30m) or more were not uncommon. Some of the earliest scrolls were discovered by chance in 1947 in a remote area near the Dead Sea, and are now known as the Dead Sea Scrolls. They were preserved in large earthenware jars and certainly date from the AD 100s, perhaps even earlier. They are of great value to scholars for they contain parts of the books of the Old Testament written in Hebrew.

By the AD 300s the need to produce documents rapidly led to the use of rounded letters in the monasteries, the main source of manuscript books. By the end of the 700s the monastery of Saint-Martin, at Tours in France, became a great centre of writing with over 200 monks at work. They devised a script which was used for more than 600 years until printing was invented.

The monks of the Middle Ages used quill feathers from swans or turkeys for writing, and these were the main pens in use until the 1800s. In 1803 a steel nib was invented, and by 1830 these nibs were being produced in Birmingham, England, by machine. In the 1880s the fountain pen was invented by Lewis Edson Waterman of New York City. The earliest of these pens had a reservoir in the stem which had to be filled with an eye-dropper. The self-filling pen followed in the first few years of the 1900s. It had a lever on the shaft. The nib of the pen was dipped in ink, and the lever was raised then lowered to deflate and inflate the rubber reservoir in the shaft. The pen was then full.

Also in the United States, John J Loud patented a ballpoint pen as early as 1888, but it was not a commercial success. The modern ballpoint pen was invented in the 1930s by two Hungarian brothers, Georg and Ladislao Biro, but it was not freely available until the mid-1940s when a factory for making the pens was established in England. The ballpoint pen is now produced so cheaply that it has become a throw-away item.

The traditional alphabet used for writing has been used as a basis for various codes designed for rapid visual communication. A semaphore consisting of a series of discs and shutters was devised in 1767 by an English inventor, Richard Lovell Edgeworth, who was anxious to know the result of a horse race at Newmarket. Claude Chappé used an improved system in 1794 to send messages from Paris to the French frontier armies near Lille.

Various systems were used by the British Admiralty in the 1800s, but it was not until 1890 that actual words were spelt out, using a semaphore alphabet

Book maker

Rivalry between two monarchs led to one of the most revolutionary inventions in the history of communications — the codex, or multi-leaved book. About 5000 years ago, scribes made books from papyrus sheets, stuck together to make a long roll. Ancient libraries contained thousands of these scrolls.

Around 190 BC the Pharaoh Ptolemy V of Egypt found that King Eumenes II of Pergamum in Asia Minor was collecting a library that rivalled his own — so he cut off supplies of papyrus to Pergamum. Eumenes' experts thereupon produced a substitute made from leather, known ever since as *parchment*, meaning 'material from Pergamum'.

The new material was stronger than papyrus; moreover, it could be sewn, and scribes could write on both sides of it. Soon sewn books were made — and after 400 years they completely ousted the old scrolls.

based on the relative positions of two arms, either human or mechanical. The semaphore was very efficient in fine weather, and could be seen from a considerable distance with a telescope, but it was useless in poor visibility.

The use of flags for signalling was first introduced in the 1600s, but it took over 200 years to devise an efficient system. However, one invented by Sir Home Riggs Popham was adopted and it was his code that was used at Trafalgar to signal Nelson's famous 'England expects' message. The code was simplified in 1817 by Frederick Marryat – more famous as a writer of adventure stories – and was later adopted as the official code for flag signalling.

Navies usually have their own codes, but in 1948 an international alphabet was adopted for use by merchant shipping. It has a flag for each letter of the alphabet; some letters used alone have a special meaning. G, for example, means 'I want a pilot'; V means 'I require assistance'.

Unfortunately, the blind cannot use visual communication. This problem particularly concerned Valentin Hauy, a French professor who realized that the sense of touch is highly developed in blind people. According to tradition, he impressed writing heavily on a piece of paper, and one of his pupils succeeded in reading it. He then started to print letters in relief and in 1784 produced a book for the blind. Institutions for the blind were set up in Paris, St Petersburg (Leningrad), and Liverpool, and the movement gradually spread elsewhere.

Hauy's system was later refined by Louis Braille, who was blinded as a child by an accident in his father's saddler's shop. He spent many years deciphering embossed messages, and from what he had learned he produced in 1829, a six-dot code which later became the standard system in France. Years passed, however, before it was widely used. It was due to the enthusiasm of Dr T R Armitage, a London doctor, that it was generally adopted after the foundation of Britain's National Institute for the Blind in 1869. The institute has published books and pamphlets in Braille ever since.

Printing originated in China, and the first printed book to have survived from that country dates from AD 868. The Chinese also discovered that engraved

wooden blocks might be used to print sheets which could be bound together.

Sometime between 1040 and 1050 came the invention of printing from movable type. It is attributed to a Chinese scientist, Pi Sheng, and he made his type from pottery. About 300 years later, wooden type was also invented in China. But these

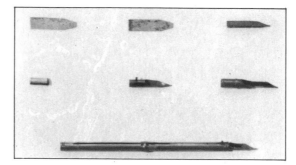

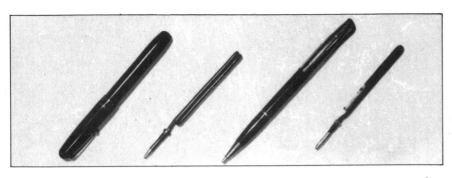

Pen portraits through the years:
Far left: early steel pens, patented in 1808, with above them the 'Penographic' fountain pen of 1819 invented by John Scheffer in Britain. *Near left* is a Parker pen of about 1900. *Right,* the first lever-filled pen, invented by Walter Sheaffer in 1908. *Above,* the ball point pen, patented by Laszlo Biro in 1938.

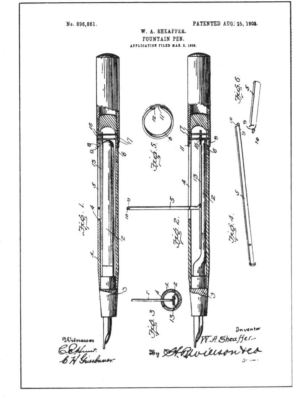

Far left: Semaphore signalling with flags came into military use during the 1800s.
Left: The use of Morse quickly spread to all countries and languages. This early operator was a Frenchman.

Far left: Braille, invented in the 1820s, has brought the pleasures of reading to millions of blind people.
Left: Claude Chappé's semaphore of 1794.

Methods of printing changed little for 300 years, though various improvements were made and new designs were produced for typefaces. In the 1700s John Baskerville entered the printing trade in Birmingham and worked on the improvement of inks. He devised a method of smoothing paper by pressing it between hot copper plates, and designed a typeface used first in 1757. Baskerville type is still much admired.

In the 1700s most paper was made by hand from linen and cotton, but experiments proved that it could be made from wood. A Frenchman, Nicolas-Louis Robert, eventually perfected a paper-making machine for which he was granted a patent in 1798. It had a continuous belt of wire mesh which revolved over a frame. Wood pulp was spread over the wire, which was vibrated to shake off excess water. The pulp was then squeezed through rollers and dried. The patent was later acquired by the British papermakers, Henry and Sealy Fourdrinier, who patented their own continuous papermaking machine in 1807.

The first iron printing press was invented in 1798 by Charles, third Earl of Stanhope, with the help of Robert Walker of Vine Street, London, an experienced engineer. In this press a much greater pressure could be used than in the earlier wooden presses.

The coming of steam-power brought great changes in printing in the early 1800s. Steam was used in the United States to operate the traditional bed-and-platen presses, by Daniel Treadwell in 1822 and eight years later by Isaac Adams, both of Boston. However, a speedier process was needed if newspapers were to be printed in quantity daily.

The solution was a steam-driven cylinder press, invented by Friederich Koenig, who was born in 1774 in Saxony. The cylinder was used to press the paper on to the type. Koenig moved to London and set up a works to manufacture the new machines, and in 1812 demonstrated that a cylinder machine could take off impressions at the rate of over 1000 an hour. On 28 November, 1814, one of his presses was used to print *The Times* newspaper. Improvements followed: in 1846 the Philadelphia *Public Ledger* was printed on a machine produced by the New York firm of Richard M Hoe. This had revolving type locked in place on a cylinder which revolved at considerable speed. Each cylinder printed 2000 sheets an hour and, since machines could be fitted with up to ten cylinders, the output was very great. Between 1856 and 1862 several British newspapers acquired Hoe machines.

Meanwhile, experiments were proceeding in France to try to find ways of avoiding the necessity of locking up large amounts of type on the cylinders. In 1829 Claude Gennoux successfully made moulds by pressing layers of wet paper against the type. After drying, these moulds – or matrixes – were used to cast metal plates. Twenty years later a Paris printer, Jacob Warms, succeeded in making curved plates which could be used on a cylinder press.

By 1861 New York newspapers were using curved plates on their rotary printing presses for the first time, and in 1882 a New York paper was using a rotary press to print 24,000 12-page papers an hour. This speed of production was made possible by feeding the paper into the presses from a roll in a continuous sheet, called a web, instead of in separate sheets. Another American, William Bullock, invented the first web-fed press in 1865. The press incorporated machinery for cutting the paper into separate sheets as it was printed, and within five years Bullock and Hoe had devised a machine which folded the papers as well.

discoveries did not penetrate the world outside China.

In the West the earliest printing was done in the late 1300s, using engraved wooden blocks, but the main developments took place in the mid-1400s when Johann Gänsefleish Gutenberg, a goldsmith of Mainz on the Rhine in Germany, reinvented movable type. Gutenberg found that it was possible to mould small pieces of a lead alloy, each carrying a single letter, and to assemble these by hand to form words and sentences.

Between 1452 and 1455 he produced an edition of the Bible in two volumes, using a printing press of the bed-and-platen type not unlike the wine press of the period, and possibly adapted from just such a press. His Bible had in all 1284 pages, set in two columns of type each of 42 lines, the whole involving the setting of 3,000,000 type characters by hand.

Several pages of type were set up and printed. Then the type characters were used again to set up a few more pages, and the process was continued until the whole book had been printed and the sheets could be bound together. Printing quickly spread over the rest of Europe. It brought books within the range of all who could read, and was the biggest single factor in the spread of learning and ideas. It contributed to the revival of scholarship now called the **Renaissance** and, by giving more people a chance to study the scriptures, was a major factor in the Reformation.

Far left: Johannes Gutenberg, who invented printing in Europe in the 1400s.
Left: Two columns from the Mainz Psalter of 1457, printed from type made by Gutenberg; the actual printing was probably the work of Gutenberg's business partners Johann Fust and Peter Schöffer.

One of the problems resulting from the increase in speed was that compositors setting up the type by hand could not keep pace with the machines. Effectively, typesetting had changed hardly at all since Gutenberg's day. Several attempts were made to invent typesetting machines in the mid-1800s, and the first successful one was devised simultaneously by two Americans – Robert Hattersley, in 1866, and Charles Kastenbein, three years later. These machines used ordinary type. A Kastenbein machine was used by *The Times* from 1872 to 1905.

The real breakthrough came in 1886 when a German immigrant to Baltimore, Maryland, Ottmar Merganthaler, produced the Linotype machine. Merganthaler solved the problem by casting the type as required in solid lines. The operator of a Linotype uses a keyboard similar to that of a typewriter, and as he depresses each key a brass matrix for that particular letter drops into place. When the line is complete, the row of matrixes is placed over a mould and the line of type is cast, the molten lead alloy setting almost immediately. Merganthaler was enabled to devise his machine because of another invention – a machine for mass-producing hard steel punches with which the brass matrixes could be made.

A similar principle is used by the Monotype machine, invented in 1889 by another American, Tolbert Lanston. This machine is in two parts. The first, operated by the keyboard, punches coded holes in a paper tape. The second has the tape fed into it, and the code controls the casting operation. In Monotype casting, however, every letter and space is cast separately, just like the type produced so laboriously by Gutenberg 500 years ago — but in the right order.

Gutenberg's system of printing, still the main one in use today, works by the application of ink, and subsequently paper, to a raised surface; it is known as *relief printing*. But two other methods of printing have also been evolved. In *planographic printing* the design to be printed and its background are one flat surface; in *intaglio printing* the part to be printed is etched or cut into the plate – the exact reverse of relief printing.

Planographic printing owes its invention to the ambition of Aloysius Senefelder, a Bavarian actor and playwright who was anxious to publish his own works. Artists were already making prints by hand from slabs of stone, etched with nitric acid. Senefelder found that some kinds of stone absorb both oil and water. He drew on the stone with a greasy crayon, and then dampened the stone, which absorbed the water only where there was no crayon design. He then made an ink of wax, soap and lampblack which stuck to the crayon and came off on paper, producing a print. The term lithography comes from the Greek word for stone, *lithos*.

Later inventors have replaced the stone by sheets of zinc or aluminium, and the design to be printed is applied to the plates by a photographic process. Offset lithography, the main process used today, is the result of another invention which came about by accident. An American printer, Ira W Rubel, had allowed the rubber covering of the impression cylinder to become inked, and found that it not only transferred a perfect impression to a sheet of paper, but also a rather better one than the direct contact with the plate.

The third printing method, gravure, also comes from an old artistic method, in which a design is cut into the surface of a printing plate. The plate is inked and the surface is wiped, leaving ink only in the engraved lines. Engraving by photography, known as photogravure, was invented by the British photographic pioneer William Henry Fox Talbot in 1852. The half-tone process, by which light and shade are represented by dots of varying size and closeness, was invented by the combined work of several experimenters in the late 1800s; the final commercial success was the work of the brothers Max and Louis Levy in 1893. The half-tone process is used in relief and gravure printing.

The development of printing methods dependent largely on photography – litho and gravure – meant that metal type was no longer a necessity. William Friese-Greene, the British photographic pioneer, patented a system of photocomposition in 1895, but it was not until 1947 that the first machine to 'set' photographically was invented, the work of engineers of the Intertype Company. Since then many developments have taken place; all of them, like so many modern inventions, the work not of one man but many. Now a typist can operate a keyboard and the machine will automatically break the words up into lines of the right length, justify them – that is, space them out so that they make full lines – and also break the words in the right place if they turn from one line to another. Modern photocomposition is computer-controlled.

A 16th-century typefounder at work. Every letter had to be individually cast in a hand-held mould.

HISTORY OF PRINTING

Above: A printing shop in the 16th century was not so very different from one 300 years later: type was hand-set and presses were hand-operated.
Above right: The oldest known illustration of a printing press is German, and dates from 1507.
Right: Great strides were made during the 19th century, as can be seen from the typefounders at work around 1800, and, *far right,* Charles Kastenbein's automatic typesetting machine which was invented in 1869.

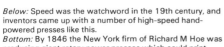

Below: Speed was the watchword in the 19th century, and inventors came up with a number of high-speed hand-powered presses like this.
Bottom: By 1846 the New York firm of Richard M Hoe was producing giant rotary power presses which could print several thousand sheets an hour.
Right: By 1851 Aloysius Senefelder's lithographic process had brought a new style of printing press into existence.
Bottom right: A modern multi-unit web press can print, fold, and deliver several thousand copies an hour of full-colour material.

Telegraph and Telephone

The possibility that one day Man might succeed in communicating over long distances exercised the minds of many inventors for hundreds of years. The first experiments with the electric telegraph began in the early 1700s, but there were then no sources of adequate electric current available to allow a signal to be sent over a significant distance. Even so, experimenters in London managed to send a message over 2 miles (4.2 km) before 1730.

The opportunity to advance came in 1800 with the invention of the electric battery by the Italian scientist Alessandro Volta, who had already envisaged the installation of overground wires for electric signalling between the Italian cities of Milan and Como. Volta's battery meant that for the first time experimenters had sufficient current to transmit signals.

In 1819, the Danish physicist Hans Christian Oersted found he could use an electric current to deflect a magnetic needle, and that the direction of the movement depended on the direction of flow of the current.

The next stage in development came in London, where in 1837 a partnership was formed between two English scientists, William Fothergill Cooke and Charles Wheatstone. After many experiments, the two men took out a patent for the first true electric telegraph in 1845. It was a single-needle apparatus in which a key was depressed to connect a line with a battery. The current travelled to another key at the far end of the line and through a recorder to earth, deflecting the recorder needle.

Meanwhile an American artist and sculptor, Samuel Finlay Breese Morse, had become interested in telegraphy. In 1832, on a voyage from London to New York aboard the sailing ship *Sully*, Morse met a Dr Jackson, who told him about the fascinating new discoveries and the problems of passing messages by

wire. Thereafter, Morse's artistic work took second place in his life. He did not rest until, in 1837, he had produced a working model of a telegraph.

Morse's first telegraph depended on the making and breaking of an electric current on the telegraph line; the signals were recorded at the end of the line by a pen touching a moving strip of paper. The movement of the paper was governed by a magnet above the pen. A short signal was a 'dot' and a longer signal a 'dash'. There was a short interval between letters and a longer interval between words. Morse devised a code to cover the letters of the alphabet and the numbers 1 to 10 – a code still in use.

The early type of hand-operated telegraph was slow. Inventors brought many improvements. The partnership of Cooke and Wheatstone devised a system in which the message was recorded by punching holes in a strip of paper.

The strip was then passed through the transmitting apparatus, and by this method a speed of up to 400 words a minute was achieved. It had the great advantage that the same message could be transmitted a number of times from the same strip of paper.

By 1845 attempts were being made to link the great land areas of the world by laying telegraph cables on the sea bed. In 1850 a cable was laid across the English Channel. In 1856 the Atlantic Telegraph Company was formed with the idea of linking Britain with America. This involved the co-operation of inventors working in different fields. The chief engineer was the Englishman Charles Tilston Bright and, although early attempts failed, he persisted and perfected machinery for paying out cable from a ship.

The successful operation of the cable owed much to the British electrical engineer William Thomson, who later became Lord Kelvin. He devised a small and delicate instrument, the mirror galvanometer, which could measure the weakest electrical current. Another inventor worked out a formula for making an adhesive mixture of resin, gutta-percha and tar which was not only waterproof but was also a good insulator. Chatterton's compound, as it was called, was used in the manufacture of the 1896-mile cable which was eventually laid between Ireland and Newfoundland. The first telegraphic message was passed across the Atlantic in 1858.

One of the British consultants used by the Atlantic Telegraph Company was Cromwell Fleetwood Varley. He invented an intricate machine, a chronopher, which was installed at London's Greenwich Observatory. It was able to transmit Greenwich Mean Time by telegraph to all parts of Britain twice daily, at 10am and 2pm.

In America, Thomas Alva Edison entered the field, and by 1872 he had perfected duplex telegraphy, which made it possible to transmit two messages simultaneously over the same line. The quadruplex system and the five-unit single code followed and, later, the tape machine or teleprinter, in which the message is transmitted and received automatically.

Telegraphy can also be used to transmit still pictures. The picture is scanned by a light spot which interprets the various tones in terms of variations in amplitude of an alternating current. This passes by line to the receiver, where a light scans the surface of photographic paper, reproducing the original picture.

Shelford Bidwell, an English barrister and physicist, devised a picture-transmission system in 1881, and by 1907 the first complete equipment had been assembled by French engineers. The system was perfected during the 1920s and 1930s by the firms of Siemens in Germany and Belin in France. After World War Two further inventions in the system were made

Above, an early use for the electric telegraph — a stock ticker, which kept investors abreast of market movements.
Left: A reconstruction of Samuel Morse's first design for a telegraph apparatus.

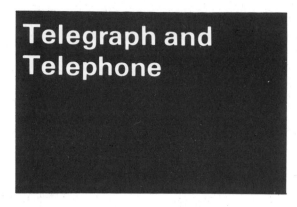

The development of satellites in the 1960s and 1970s has revolutionized communication. This Intelsat IVa global communication satellite of 1975 is typical of the systems now in use.

Pictures by wire are almost as old as telegraphy itself. The early systems used a scanning pen at one end on a metal original, with varnish protecting the unwanted parts, and another pen acting on chemically-treated paper at the other end. The picture *above* was drawn at Marseilles and wired to Paris in 1863.
Above right, a Belin transmitter-receiver of 1949.

and caused changes in the magnetic field. The electric currents produced in the coil corresponded to the sounds. In 1876 Bell succeeded in passing a vocal message to an assistant in another room. To link telephones together, some form of switching system was necessary, and engineers devised the first switchboard and installed it in New Haven, Connecticut, early in 1878.

The idea of a telephone which linked callers visually as well as aurally was a dream of inventors for many years. The breakthrough came in the early 1970s, when the Bell Telephone Company of the United States introduced a video-telephone service, incorporating a TV-type screen.

by Fred Jarvis of the London *Daily Mail*, and machines were manufactured in England.

The suggestion that people might actually be able to speak to one another by wire was put forward by a number of scientists, but credit for the invention of the telephone goes to Alexander Graham Bell, a Scot. Bell was fortunate enough in his early days to know Charles Wheatstone, who had invented the telegraph, and Alexander John Ellis, who was an expert on sound. Ellis showed Bell that the vibration of a tuning-fork could be influenced by an electric current. He was able to produce sounds not unlike those of the human voice.

In 1872 Bell went to teach deaf-mutes in Boston, Massachusetts. In his spare time he worked on a musical telegraph. After studying the human ear drum, he constructed similar artificial drums from thin sheets of metal and linked these with electric wire. He then constructed more sophisticated apparatus with metal diaphragms suspended close to coils which could be magnetized.

The sounds reaching the diaphragm made it vibrate

News flash

Politics was responsible for getting Samuel Morse's invention of telegraphy accepted. Morse could not get financial backing from commercial sources, but the US Congress voted to build a test line between Washington, DC, and Baltimore, Maryland.

On May 24, 1844, Morse sent over the line from Washington a solemn message: 'What hath God wrought'.

Still there was no excitement — until a few months later, when news of James K Polk's nomination as Democratic presidential candidate was flashed along the line from Baltimore to Washington. At once the US newspapers realised what they were missing, and soon a spider's web of telegraph wires was being strung across the entire United States.

HISTORY OF TELEPHONES

Two pioneer telephones — *above*, the Gower-Bell apparatus of 1879, and *left*, the first wall-mounted instrument used in the British telephone service.

Above: Alexander Graham Bell's first telephone, the 'gallows frame' model he constructed in 1875. It failed under test and, because of pressure of other work, Bell postponed further experiments for a year. But he patented his idea.

Below: Bell's magnetic telephone, the result of a year's hard experiment by Bell and his assistant, Thomas Watson. The apparatus came into commercial use early in 1877.

Two more historic telephones: *right,* a table model telephone of about 1900; to call the operator you cranked the handle at the side; *far right,* the first dial telephones were installed in New Jersey, in the United States, in 1914.

From wires to buttons: *above,* an apparently hand-knitted private exchange of the late 1880s; *right:* a switchboard at the Royal Exchange, Manchester, England, in 1895; *far right,* top: by contrast, today's telephone operator is strictly press-button orientated, and so is the latest model provided for subscribers, *far right, bottom.*

Broadcasting and Recording

The first man to send a message without using wires was James Bowman Lindsay, who used water to conduct an electric current and claimed the invention of wireless telegraphy in 1853. But it was many years before messages were carried by radio waves.

In the 1870s a Scottish scientist, James Clerk Maxwell, who had become the first professor of experimental physics at Cambridge University, argued that wireless telegraphy would be possible by using electro-magnetic waves, but he did not offer any practical demonstration to back up his ideas.

In 1885 a Welsh electrical engineer, Sir William Preece, sent currents between two insulated squares of wire a quarter of a mile (0.4 km) apart. Then in 1887, Heinrich Rudolf Hertz, a German physicist, proved the existence of radio waves, which could be reflected from conducting objects, such as metal sheets, in the same way as light is reflected from a mirror. He measured the velocity of these waves and opened up a vast new field for experiment.

Soon after the death of Hertz in 1894, an English physicist, Sir Oliver Lodge, gave a demonstration to the British Royal Society which proved that messages could be transmitted and received without wires. Three years later he invented a way of tuning the transmitting and receiving circuits by adjusting coils of wire. In this way the radio waves could be restricted to a definite wavelength.

Lodge's work in the 1880s had been paralleled by the work of Guglielmo Marconi, the son of an Italian father and an Irish mother. Marconi was brought up and educated in Bologna, where he started his experiments in a laboratory in the upper room of a country house. He was greatly helped by the invention, by a French electrician, Edouard Branly, of the coherer, which is basically a small glass tube loosely filled with iron filings. When electro-magnetic waves are passed through the tube, the metal particles cohere and form a conductor.

Marconi improved the Branly coherer. He replaced iron filings by nickel and silver filings, and fitted silver plugs to close the tube. He also devised an automatic tapper to shake the filings apart after a wave impulse had been passed. Within a short time he had discovered a method of transmitting signals without wires.

Marconi first succeeded in passing signals across a room, using a transmitter and receiver. Then he increased the distance by fitting both transmitter and receiver with aerial and earth. Because little interest was shown in his experiments, he decided to leave Italy and to work in London, where he was encouraged by Preece, who was then engineer-in-chief of the Post Office telegraphs. In 1896 Marconi filed an application for a wireless patent in London. He was soon sending long-distance messages by morse code,

first across the Bristol Channel and then across the English Channel.

Marconi's new wireless telegraph was used to save a ship in distress in the North Sea, and the practical advantages of the invention for shipping were fully realised after two passenger-ships had collided in a fog off the east coast of America. One of these ships had a Marconi wireless-operator who radioed for assistance and, as a result, more than 1500 passengers were saved. Radio equipment became essential on all ships and the British Admiralty paid Marconi £20,000 a year for the use of his system in the British Navy.

It was assumed by many sceptics that the use of wireless would be restricted by the curvature of the Earth and that radio waves, travelling as they thought in straight lines, would end up in space. Marconi, however, was convinced that it would be possible to send messages from Britain to America. A radio station was set up at Poldhu in Cornwall with 20 masts, each over 200 feet (60m) high.

In 1901 Marconi went to Newfoundland, used a kite to lift an aerial wire to an even greater height, and for two days received messages from Cornwall. Following that success, Marconi installed receiving apparatus on board the liner *Philadelphia*, and kept in touch with the Poldhu station throughout a transatlantic voyage.

These messages were all tapped out in morse code. Would it be possible for the human voice to be transmitted by wireless? In 1902 R A Fissenden of the University of Pittsburgh transmitted the sound of a

Within eight years of the first transatlantic broadcast of speech, people were listening to the radio on loudspeaker sets like this model of 1923.

Above: The device which made radio a practical proposition — the thermionic valve invented in 1904 by John Ambrose Fleming.

Above, right: Marconi's historic transmitter and receiver, which were fitted with concave reflectors to transmit the radio waves in the required direction.

One of the first Marconi receivers constructed to make use of Fleming's new thermionic valve.

human voice over a distance of 1 mile (1.6km), but further progress was not possible until another invention — the thermionic valve — was introduced in 1904 by John Ambrose Fleming, an English electrical engineer. The thermionic valve, also called the electron tube, serves to change the minute alternating current of a radio signal into a direct current, capable of actuating a telephone receiver or the needle of a meter. It was called a valve because the current could flow only one way through it.

Fleming's valve was greatly improved three years later by the American physicist Lee De Forest. De Forest's valve enabled the weak radio signal to control a much stronger electrical current which fluctuated in the same way as the signal. Thus the first amplifier was invented.

In 1910 De Forest fitted up a so-called 'radio-phone' on the roof of the Metropolitan Opera House in New York. With its aid listeners could hear the voices of the singers up to 100 miles (160km) away.

World War One gave an impetus to invention. Warships fitted with wireless were deployed from London, and the Germans began to use radio for wartime propaganda. Wireless was also used to send messages to and from aircraft in flight.

The idea of sending a picture by radio followed naturally after the invention of transmitters for the human voice. In 1862 an Italian priest, the Abbé Caselli, had devised a method of sending shadows over a telegraph line — proof that the idea was possible.

The material to make the idea possible was also to hand — selenium, discovered by the Swedish chemist Jons Jakob Berzelius in 1818. In 1873 a British scientist, Willoughby Smith, found that selenium, which normally does not conduct electricity, will do so readily when exposed to light. Ten years later the American Charles E Fritts invented the first photo-electric cell, which could be used for controlling current by means of light.

In 1884 Paul Nipkow, working in Germany, devised a mechanical scanning disc which was used with a selenium cell, but the images were too faint for practical purposes. By this time, a number of scientists were working on the problems.

One further device had been invented in 1878 by the London scientist Sir William Crookes. This was the Crookes tube, a glass vessel containing very little air, through which a high voltage is passed. The action of the current on the remaining air makes the walls of the tube glow fluorescently, due to a stream of electrons emitted from one of the electrical terminals, the cathode. Here was the first cathode-ray tube, ancestor of all TV tubes.

In 1887 Karl Ferdinand Braun of Strasbourg produced the first cathode-ray oscilloscope, as it was called. It made the wave form of alternating currents visible to the human eye. But 10 years elapsed before the cathode-ray tube was used by Professor Boris Rosing in St Petersburg to produce a television system. In this system two mirror drums scanned the image and the light reflection was thrown on to a photo-electric cell. Unfortunately, the resulting picture was of poor quality.

Rosing's work was carried on by one of his students, Vladimir Zworykin, who went to the United States after World War One and in 1923 patented a camera tube. His pictures however were little better than Rosing's. The first successful electrical transmission of a picture was between London and Paris in 1907, using a method invented by Dr Korn of Berlin.

In Britain, Alan Archibald Campbell Swinton took out a patent for a television system in 1911, but made no great claim for it. The picture was poor. After World War One, the Scot John Logie Baird carried out a series of experiments and in 1926 he demonstrated what he called 'seeing by wireless'. The practical stage had at last been reached, and British Government approval was given for experimental television transmissions. These were begun in 1929, in

Communication

co-operation with the British Broadcasting Corporation, and continued for five years.

However, American developments based on the work of Zworykin had produced a rival system in which the picture to be transmitted was electronically scanned – Baird used a mechanical scanner – and eventually the Baird system was dropped.

At this time, further developments in the field of radio were taking place. The work of German physicist Heinrich Hertz in the 1880s on the reflection of radio waves was followed up in 1900 by Nikola Tesla, in the United States. Tesla, a prolific inventor, put forward the idea that radio waves might be used to locate ships at sea, and it was developed as 'radio-location' or 'radar'.

Distant objects are detected and their directions and ranges measured by means of radio-wave echoes reflected back to a detector apparatus. In 1936, realising the potential value of radar in wartime, the British Air Ministry asked the physicist Robert Watson-Watt and a team of scientists to study rays which might be directed at attacking aircraft to stop their flight. Watt's team made a device which detected the presence of moving aircraft by radio-magnetic waves reflected back to their source. Radar was used on a considerable scale in World War Two.

At the same time, two British scientists, Harry Boot and John Randall, invented the cavity magnetron to locate surfaced submarines in the dark. It worked on the same principle as radar, but produced very powerful electro-magnetic waves which could be directed and operated like a searchlight, sweeping rather than flooding the area to be searched.

Radio has brought far-reaching changes in the study of astronomy. In 1932 Karl Guthe Jansky, who worked at the Bell Telephone Laboratories in New Jersey, discovered that radio waves reached Earth from outer space. Unfortunately, he was unable to build the large dish-shaped aerial he needed to detect their source, but Grote Reber, a radio enthusiast of Wheaton, Illinois, took up the challenge. He built a parabolic reflector of galvanized iron and wood in one summer, stimulating great curiosity among his neighbours. Using the reflector in his spare time between 1938 and 1944, he succeeded in making radio maps of the Milky Way, and also recorded radio waves from the Sun.

As well as trying to transmit sound, inventors spent a considerable amount of time trying to record it. The idea of trying to record the human voice came to the American inventor Thomas Alva Edison while he was trying to invent a way of recording telegraph messages in Morse. He sketched out his invention and handed the plan to his assistant, John Kruesi, to work up into a machine. Kruesi, who had no idea what he was making, nearly fainted when the machine repeated Edison's words, spoken into its mouthpiece. The words? 'Mary had a little lamb'!

Edison used tinfoil wrapped around a cylinder as a recording medium. Two other Americans, Chichester A Bell and Charles S Tainter, invented a wax-coated cylinder in 1885. The disc system of recording was the invention of Emile Berliner, a Hanover-born US citizen, in 1887. Berliner also invented the so-called lateral-cut method of recording, in which the needle moves sideways instead of up and down.

The tape recorder is almost as old as the phonograph (cylinder model) and gramophone (disc model). A Danish engineer, Valdemar Poulson, invented a magnetic recorder in 1899, in which messages were recorded on steel tape. Experiments with steel wire and tape continued for many years, until two scientists invented a method of coating strips

Above: How it all began — John Logie Baird's original TV apparatus of 1925. *Left:* As it is today — sophisticated modern cameras in action in a studio, while sensitive boom microphones pick up the speakers' voices.

The latest use of television: The British Broadcasting Company's 'Ceefax' system transmits any one of 100 pages of news and information, and the viewer selects which pictures he wants by dialling a code number.

Edison's first phonograph: it was on this machine, or one like it, that he made the first recording, of the words 'Mary had a little lamb . . .'

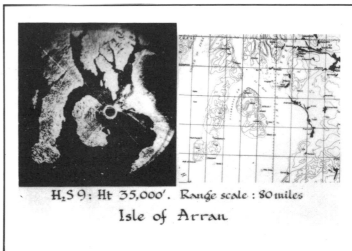

H₂S 9: Ht 35,000'. Range scale : 80 miles
Isle of Arran

The story of radar: *top left,* beacon towers like this helped to guard Britain against enemy air attack during World War Two; *top right,* a modern high-powered multi-beam surveillance radar set fitted in the cockpit of a Lancaster bomber during World War Two; *bottom right,* a radar trace with a map of the area scanned from the air.

of paper with magnetic materials. They were J A O'Neill in the United States in 1927, and Fritz Pfleumer in Germany in 1928.

The development of plastics led to the use of cellulose acetate tape in Germany during the 1930s, and after World War Two plastic tapes were adopted everywhere. Machines that could record television material were developed in the 1950s and 1960s, the work of many inventors. Now in schools and homes having videotape recorders, TV transmissions can be replayed at will through ordinary domestic sets.

Sophisticated modern developments: *below,* a push-button colour TV set; *right,* a stereo record-player.

Too Soon

The worst fate that can befall an inventor is to make his invention too soon. That was the fate of the German engineer Christian Hülsmeyer.

Back in 1904, he took out a series of patents for a radio echo device to prevent ships from colliding . . . an early form of radar. No-one was interested.

When World War Two broke out in 1939, both sides had some form of radar; but the Germans were left badly behind in the development of the device. Had they followed up Hülsmeyer's idea in 1904, they would have been well ahead of the British who were saved by radar in the air battles of 1940.

Photography and Cinema

William Fox Talbot's photograph of a window at his home, Lacock Abbey, in Wiltshire. The exposure took several hours.

For the ancestry of the camera and the cinema we have to go back to the ancient Greeks, to the philosopher Aristotle who was tutor to the future Alexander the Great in the 300s BC, and whose works were considered incontrovertible by European philosophers until the AD 1500s. Among the many things described by Aristotle in his works was a device we now know as the camera obscura.

The camera obscura is a darkened room in which an image of the landscape outside it is projected on to a screen mounted on one interior wall. The image comes through a tiny hole in the opposite wall. A camera obscura of this type was used by the Arabian mathematician Al-Hasen in the late AD 900s.

The use of a lens instead of a tiny aperture was probably the invention of the Neapolitan scientist Giovanni Battista della Porta, in 1568, though it had been suggested by a Milanese doctor, Girolamo Cardano, nearly 20 years before. By the 1700s the camera obscura was a device commonly used by artists.

The next crucial step in the history of photography came in 1725 when a German anatomy professor, Johann Heinrich Schulze, found that silver salts darkened on exposure to light. Schulze coated some paper with silver salts, and with this made stencil-like images which, however, rapidly darkened completely on further exposure.

The idea of combining the camera obscura and light-sensitive paper to produce a picture occurred to several people around the end of the 1700s and in the early 1800s. The first person to succeed was a retired French army officer, Joseph Nicéphore Niepce. In 1826 Niepce exposed a metal plate coated with a layer of bitumen to the image in a camera obscura. The light hardened the bitumen, but where the light had struck it the layer of bitumen could be washed away. Niepce then etched the uncovered portions, and called the result a 'heliograph'. It was a negative image: the highlights showed dark and shadows light.

Niepce continued to experiment, and produced the first true camera. It consisted of two wooden boxes, one carrying a lens and the other a ground-glass screen. The boxes were connected by bellows so that the distance between the lens and screen could be varied. He also invented an iris diaphragm which could be adjusted to vary the size of the aperture and thus sharpen the image.

Meanwhile a French theatrical designer, Louis Jacques Mandé Daguerre, had been experimenting in the same field. In 1830 Daguerre and Niepce decided to work together, but three years later Niepce died.

Daguerre carried on alone and eventually succeeded in fixing images on metal plates coated with silver iodide, which he treated with mercury vapour in a darkroom. These images became known as daguerrotypes. They must be viewed at exactly the

right angle in relation to the source of light if the viewer is to see a positive picture.

For some years, daguerrotypes were extensively used for portrait photography, and engravers often copied them to produce the illustrations for books and journals. Many people objected to the new invention on religious grounds; one German newspaper, the *Leipziger Stadtanzeiger,* commented: 'God created man in His own image, and no man-made machine may fix the image of God'.

Other scientists followed up Daguerre's lead and produced their own versions of the photographic process in the 1830s, many of them using paper instead of Daguerre's metal plates. Photography as we know it today derives not from Daguerre's work but from that of an English landowner and scientist, William Henry Fox Talbot, who lived on an estate called Lacock Abbey, in Wiltshire.

Talbot began his work in 1834, placing leaves on sensitized paper to obtain images and fixing the images with a salt solution. He then took a photograph of one of the windows at Lacock. The image was very small, but he found that by using a lens he could count every window-pane. The image was negative.

An early Daguerrotype camera. It could take pictures with exposures of up to 20 minutes in bright sunlight.

A medieval camera obscura. As with so many other aspects of science, the early experimenters were Churchmen.

Above: The first of the modern 35 mm cameras: Leica No 1 invented in 1914.
Right: An early use both of the magic lantern and microfilm: enlarging microscopical despatches during the siege of Paris in 1870. The films were probably flown in and out of the city by carrier-pigeon.

Far right: The Praxinoscope Theatre, a moving picture device of 1880.

Far right: Thomas Edison's kinetoscope, showing the open case and the 15 m length of film inside, which the viewer saw through the hole at the top.

Five years passed before Talbot realized the full significance of this experiment. He then obtained a positive print by placing a negative image over sensitized paper and putting it in a strong light. The prints, however, were of poor quality and tended to fade, as they were fixed only by using salt. Talbot eventually found a more stable fixer in hyposulphite of soda.

Talbot's process, the calotype, was patented in Britain in 1841. In 1847 he took out a patent in the United States, a protective action which tended to hold up progress. Many photographers, especially in Germany, preferred to continue using daguerrotypes. However, in 1854 Talbot gave up his patent rights and the field was open to all.

Meanwhile in 1850 the English sculptor Frederick Scott Archer had invented the collodion, or wet-plate, process. Archer knew that a solution of guncotton in ether would produce a useful adhesive called collodion. He used collodion with potassium iodide to coat a glass plate, and when the coating had partially dried he dipped the plate in silver nitrate. The plate was exposed in the camera while still wet and was then developed and fixed.

Details of this 'wet collodion process' were published in 1851. It had great disadvantages for the outdoor photographer, who had to carry around an immense amount of equipment including a darkroom, a tent, glass plates, chemicals and dishes. Intrepid travellers anxious to photograph mountain scenery had to engage porters to carry their equipment. But the speed and sensitivity of the process made all the effort seem worth while.

A wet-process positive was obtained by bleaching the negative, which was then backed with a dark material or with varnish. The result was similar in appearance to that of a daguerrotype but had a duller surface. The photographs produced by the wet collodion process are known as ambrotypes.

In 1853 a French teacher, Adolphe Alexandre Martin, produced what were known as tintypes. These were collodion positives on thin sheet metal – usually on tin, but occasionally on iron. The results were generally poor, but by using tintypes it was possible to produce finished photographs quickly and cheaply, and the process was therefore widely used by travelling photographers.

The Great Exhibition of 1851 in London gave a stimulus to many inventions, among them the stereoscope. This device is based on the fact that a person's eyes each register a slightly different image, so that we see objects in three dimensions. The stereoscope is so designed that perspective is given by viewing two photographs, each taken from a slightly different angle, through two lenses fitted into a box which can be held to the eyes.

The stereoscope was invented by a Scottish scientist, Sir David Brewster (who also invented the kaleidoscope), and was an improvement on an earlier

version devised about 1835 by Sir Charles Wheatstone, who used mirrors to give the impression of depth.

Since they had highly reflecting surfaces, daguerrotypes were unsuitable for use in stereoscopes. Ambrotypes were preferable and could be mounted side by side on rectangular cards. The expense of producing these cards was greatly reduced when John Benjamin Dancer of Manchester invented a twin-lens camera which took two pictures simultaneously.

In 1871 an English physician, Dr Richard Leach Maddox, invented a process using a dry plate coated with a gelatine emulsion of silver bromide – still the basis of all modern photographs. Within 10 years, dry plates were being marketed commercially. They had tremendous advantages over wet plates: photographers no longer needed to carry around so much heavy equipment, since the plates could be taken back to the darkroom for development at a convenient time. Moreover, the process required shorter exposures than wet plates, so that people no longer needed to sit still for long periods to have their portraits taken. An early daguerrotype required an exposure of more than 10 minutes. With Maddox's material it needed only 1/25 of a second!

Up to that time photography was still a hobby for the specialist. Two inventions by an American manufacturer, George Eastman, brought photography within the reach of everyone. The first was

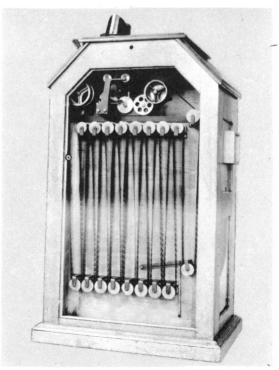

a simple snapshot camera, which Eastman called the Kodak. The camera held a roll of light-sensitive gelatin on a paper backing, capable of taking 100 pictures. When the film had been exposed, the camera had to be sent to Eastman's firm, which processed the pictures and returned the camera with a new roll of film inside it. The emulsion was transferred from the paper to glass after development.

Meanwhile, Eastman was experimenting to find a suitable film base. One answer was Celluloid, which had been invented in 1861 (see page 116). In 1888, cut film made of very thin Celluloid sheets was marketed in the United States. Eastman went one better, and in 1889 came out with an even thinner film made of nitro-cellulose. Unfortunately, this kind of film had been invented and a patent applied for two years earlier by an American Episcopalian clergyman, the Rev Hannibal Williston Goodlin. Goodwin took his case to court, saw his patent registered in 1898, and shortly after his death in 1900 his heirs were awarded $5,000,000 against the Eastman Kodak Company. Simultaneous invention can cause some awkward problems!

The lack of colour in photographs was felt to be a drawback right from the start, and many painters who had lost trade to the new science were able to win some of it back by tinting photographs. The first colour photograph was taken in 1861 by the English photographer Thomas Sutton, inventor of the single-lens reflex camera, acting under the instructions of the Scottish physicist James Clerk Maxwell. Using red, green and blue filters, Sutton took three pictures of a coloured ribbon, made transparent positives of them, and projected them on to a screen in superimposition, by means of magic lanterns with similar filters.

The Sutton-Maxwell experiment was far from perfect; not so the Photochromoscope camera invented in 1891 by Frederick Eugene Ives of Philadelphia. It took three pictures in rapid succession usung colour filters. When viewed through a viewing instrument called the Kromskop, the three pictures merged into one in full colour. A single-plate system producing a colour transparency, the Autochrome, was invented in 1904 by two French brothers, Auguste and Louis Lumière, and this was the first commercially successful system for amateurs. A three-layer system, marketed as Kodachrome, was invented in 1935 by two American amateur scientists, Leopold Godowsky and Leopold Mannes.

Other highlights in the story of photography include the invention in 1914, by the German microscope-designer Oskar Barnack, of the Leica, the first commercially successful miniature camera to use 35mm cine-film. Because of World War One, it could not be marketed until 1924. Electronic flash was invented in 1931 by the American photographer Harold E Edgerton, who specialized in high-speed photography. And in 1947 an American, Edwin H Land, invented the Polaroid Land camera, which takes a photograph and also processes it within seconds.

The use of film for photography led fairly quickly to cinematography, for many attempts had already been made in the early 1800s to produce 'moving pictures'. The experiments were all based on the phenomenon of persistence of vision, which was described in 1824 by Peter Mark Roget, professor of physiology at the Royal Institution in London, and famous today for his *Thesaurus*.

One of the earliest moving-picture inventions was the 'Thaumatropical Amusement' which appeared in 1826. Invented by an English scientist, Henry Fitton, it consisted of a small round box. Inside were a number of discs made of card with a design drawn on each.

The inventor C G Norton with his son operating an early cine-projector.

The 'Mutograph', a pioneer cine-camera, photographing the Pennsylvania Limited travelling at 100 kph.

When the discs were twirled round, the images merged and gave the impression of a single moving image. In 1833 came the Phenakistiscope, invented by Joseph Plateau, a Belgian scientist, which involved the rotation of a circular design opposite a mirror.

A variation of the Phenakistiscope, and perhaps the most spectacular of all, was the Zoetrope, or Wheel of Life, invented by Pierre Desvignes of France in 1860. The Zoetrope consisted of a drum with vertical slots regularly placed around the sides. The pictures were on a long strip of paper fitted inside the drum. The viewer peered through the slots as the drum revolved and saw figures leaping or creatures galloping through strange landscapes. He was liable to be made dizzy by the speeding images. Another Frenchman, Emile Regnaud, overcame that disadvantage in his Praxinoscope of 1877; it had no slots and the images were reflected by mirrors.

Another device to produce animated pictures involved projection. This was the Projecting Phenakistiscope, which used a revolving disc and a single-blade shutter to project animated images from drawings on to a screen.

The main impetus to cinematographic invention, however, came from a different source. In the 1870s the English photographer Eadweard Muybridge, originally Edward Muggeridge, made an attempt to take photographs of animals in movement. Working in the United States, he hoped to resolve the long-standing argument between artists and horsemen about the way in which a horse moves its legs when galloping.

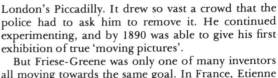

A Western Electric cine-projector of 1929 which could play back either sound film, or sound on disc synchronized with the film.

Above, a modern cine-projector in which the upper part projects the film and the lower part plays the sound track.

Far right: The 'cakestand' apparatus used in most modern cinemas. A whole programme can be laced up as one continuous length of film, with automatic cues for curtains and lights; as the film is projected it is automatically wound back ready for the next showing.

From top to bottom: a conventional wide-screen projector in action; the three-projector Cinerama system; and the Cinerama single-lens system producing the same effect.

London's Piccadilly. It drew so vast a crowd that the police had to ask him to remove it. He continued experimenting, and by 1890 was able to give his first exhibition of true 'moving pictures'.

But Friese-Greene was only one of many inventors all moving towards the same goal. In France, Etienne Jules Marey invented a photographic 'gun' in 1882 to take pictures of birds in flight, and followed it up with a camera which could take 60 pictures a second on a paper-based film. Several other French inventors took out patents for both cameras and projectors.

In 1889, using Eastman's flexible film, the American inventor Thomas Alva Edison made a kinetograph to take moving pictures, and a kinetoscope to show them. The kinetoscope was a box about 4 feet (1.2m) high with a peephole through which a viewer could look. It displayed 50 feet (15m) of film, and ran for about 13 seconds. Later the film was slowed down to last nearly 40 seconds. Edison set up 'kinetoscope parlours' in which people could view the films by putting a coin in a slot.

Successful projectors for films were invented simultaneously in 1895 by Thomas Armat and Woodville Latham of the United States, the Lumière

Using exposures of less than one-thousandth of a second, Muybridge lined up a series of cameras spaced evenly along a track, and set up strings which would operate the shutters as a horse galloped past. The photographs, viewed in a Zoetrope, revealed information about locomotion which could not be accurately noted by the naked eye.

The next step was to find a way in which such a succession of photographs could be projected on to a screen. At a meeting of the Royal Photographic Society in London in 1885, the English photographer William Friese-Greene set up a small slide projector in which the usual slide carrier had been replaced by a glass disc bearing a ring of pictures. When the disc was revolved, the audience saw for a few seconds a moving picture.

Friese-Greene perfected this apparatus and demonstrated it in the window of his studio in

CONVENTIONAL WIDE SCREEN

CINERAMA

ULTRA-CINERAMA

Brothers in France, and Robert W Paul in England. From then on, progress was largely a matter of development. The first films were silent, but Edison always planned to link his films with the phonograph he had invented. Such a link proved inefficient in practice, and the next breakthrough was the invention of sound recording on film, devised by the American radio pioneer Lee De Forest in 1922. Meanwhile another American, Herbert T Kalmus, had perfected a colour system for motion pictures, to which he gave the name Technicolor.

Later inventions were aimed at making the viewing of films more exciting. The American Fred Walter developed Cinerama, which was introduced in 1952. This involved using three cameras side by side to span an extensive area. A year later CinemaScope, based on research by the Frenchman Henri Chrétien, appeared. A 'distort' lens on the camera reduced a wide picture to the normal 35mm width; a compensating lens on the projector restored the picture to fill a very wide screen. Both systems had sound tracks and loudspeakers for stereophonic sound. In 1955 the US film producer Michael Todd introduced Todd-AO, a new system of wide-screen film, invented for him by Brian O'Brien. It used film 65mm wide. Many variations of these systems were developed in the 1950s and 1960s. But the colossal fall in cinema audiences as TV gained ground slowed invention in the cinema world.

HISTORY OF RAZORS

Although we think of Early Man as a shambling, long-haired, bearded creature, evidence found by archaeologists suggests that he probably shaved his beard. In short, the razor was one of Man's earliest inventions, antedating the wheel.

Cave drawings indicate that men used sharpened flints and pieces of shell to shave with as much as 20,000 years ago. As soon as metal-working came in, so did metal razors. Bronze, iron and even gold razors have been discovered in ancient graves, though a gold razor was almost certainly a status symbol rather than a practical tool.

There may have been good practical reasons for shaving: in battle, you could grab your enemy by the beard, and facial hair was one more thing to keep clean in an era when there were fewer facilities for cleanliness than now.

The ancient Greeks were clean-shaven. The Romans thought them cissies; but they, too, adopted the use of razors about 2400 years ago. The present steel razor evolved over the centuries, but it remained substantially unchanged in design for a long time . . . until the inventors got to work on it.

A Frenchman, Jean-Jacques Perret, invented the first safety razor in 1762. He fitted a guard along one edge to stop the blade from slicing into the skin. A similar device was made in Sheffield, England, around 1828. The present hoe-shaped pattern was evolved in the United States in 1880, but it was a fixed-blade device.

The modern safety razor was the brainchild of a US salesman, King Camp Gillette. He had been trying to think of some essential object in daily use which could be manufactured so cheaply that people would use it

and throw it away, thus creating a constant demand — and a fortune for the lucky inventor. One day in 1895 as he was having his morning shave and cursing the difficulty of keeping a razor really sharp, the idea came to him: a razor with a disposable blade.

Gillette, no metalworker, spent six years trying to make a suitable holder and a blade of paper-thin steel which would take a sharp edge. Eventually, he was introduced to a resourceful engineer named William Nickerson, who overcame all the technical problems which had plagued Gillette for so long. Even with a good product, safeguarded by patents, Gillette found his problems far from over. He needed financial backing.

So he made his razors and gave them away – and at last a wealthy friend saw the possibilities of profit and put up the money Gillette wanted. His first sale was made in 1903 – a batch of 51 razors and 168 blades. In 1908, Gillette's company produced 300,000 razors and 14,000,000 blades.

The safety razor has remained a best-seller ever since. Later refinements, introduced in the 1960s, include stainless-steel blades with a longer life, and the ribbon blade, made in a roll mounted in a throwaway razor head. As the blade becomes blunt, the roll is wound on so as to bring a fresh, sharp section into use.

The first dry-shaver was a hand-operated machine in which a moving cutter worked against a fixed one, like the two blades of a pair of scissors. It was invented in the early 1900s by an Englishman, G P Appleyard. It was only a short step to a power-driven model, and the first electric shaver was patented in 1923 by an American, Colonel Jacob Schick.

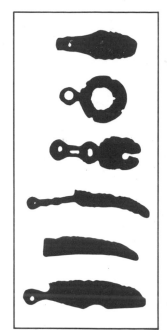

Even in Bronze Age times, men used razors. These fearsome-looking specimens were found in the River Thames near London.

"The Old Shaver."

12 MONTHS LUXURY FOR 12 PENCE
(a shilling shaving stick lasts a year.)

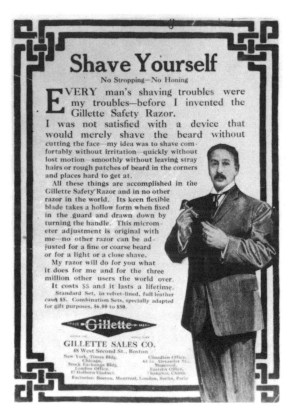

Shave Yourself
No Stropping—No Honing

EVERY man's shaving troubles were my troubles—before I invented the Gillette Safety Razor.

I was not satisfied with a device that would merely shave the beard without cutting the face—my idea was to shave comfortably without irritation—quickly without lost motion—smoothly without leaving stray hairs or rough patches of beard in the corners and places hard to get at.

All these things are accomplished in the Gillette Safety Razor and in no other razor in the world. Its keen flexible blade takes a hollow form when fixed in the guard and drawn down by turning the handle. This micrometer adjustment is original with me—no other razor can be adjusted for a fine or coarse beard or for a light or a close shave.

My razor will do for you what it does for me and for the three million other users the world over. It costs $5 and it lasts a lifetime. Standard Set, in velvet-lined, full leather case $5. Combination Sets, specially adapted for gift purposes, $6.00 to $50.

Gillette

GILLETTE SALES CO.
48 West Second St., Boston

This dry shaver worked by rocking it to and fro across the beard.

Left and far left: Shaving habits of the cut-throat era and the dawn of the safety-razor age were mirrored in the advertisements of the period.

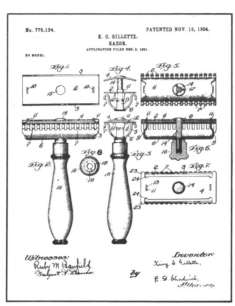

Razor development through the years: *Top right,* the old cut-throat razor, which had to be honed every time it was used; *left and right,* razors with interchangeable blades, one for each day of the week; *above,* King Gillette's patent application for his first safety razor, which he filed in the United States in 1901; *below left,* the Gillette twin-blade razor, invented in the mid-1960s; *below centre,* the first electric razor, patented by Colonel Jacob Schick in 1923; *below right,* a modern streamlined electric razor.

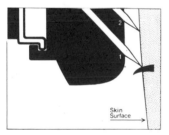

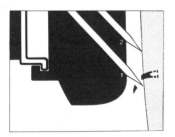

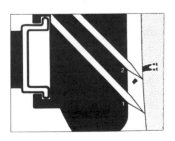

Skin Surface

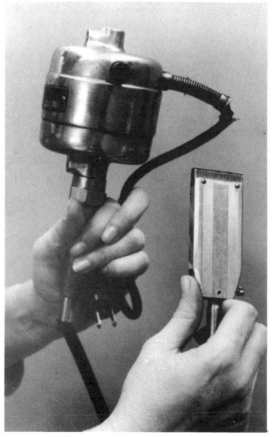

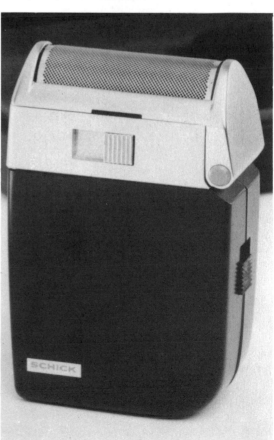

In the Home

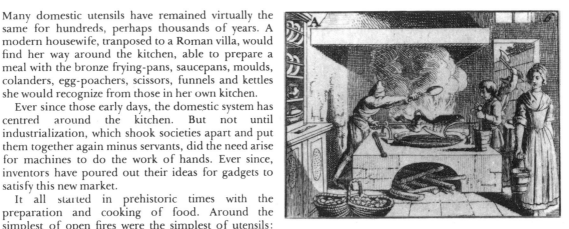

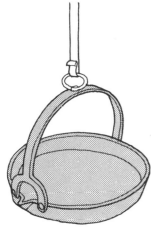

Top, an 18th-century English copper kettle with stand and spirit lamp; *centre*, a suspended frying pan of the late 18th-century; *bottom*, a 16th-century skillet.

Above right: The kitchen has come a long way since this German engraving of 1784. This version had a wood fire and a hand-turned spit.

Opposite: Probably one of the greatest inventions for the housewife was the vacuum cleaner. This was one of the first electric cleaners, in use a few years before World War One.

Many domestic utensils have remained virtually the same for hundreds, perhaps thousands of years. A modern housewife, tranposed to a Roman villa, would find her way around the kitchen, able to prepare a meal with the bronze frying-pans, saucepans, moulds, colanders, egg-poachers, scissors, funnels and kettles she would recognize from those in her own kitchen.

Ever since those early days, the domestic system has centred around the kitchen. But not until industrialization, which shook societies apart and put them together again minus servants, did the need arise for machines to do the work of hands. Ever since, inventors have poured out their ideas for gadgets to satisfy this new market.

It all started in prehistoric times with the preparation and cooking of food. Around the simplest of open fires were the simplest of utensils: stone bowls, mortar and pestle, flint blades, a makeshift spit.

One of the earliest appliances having moving parts was a grinder the women used to make flour. It consisted of two disc-shaped stones with holes in the middle. The grain was poured into the top hole and, after being ground between the stones, fell as flour through the lower hole.

The invention of pottery in prehistoric times was a great advance and led to the development of many utensils which have remained basically unchanged, like the Roman clay food-warmer with a movable bowl at the top and an oil lamp below.

Throughout the barbaric times in Europe, the central open fire dominated, with a large iron pot hanging over it on ropes, thongs or an iron bar. Inventions were still in terms of materials rather than gadgets – silver for the wealthy, horns for drinking, even some glass for bottles and cups.

In the Middle Ages, life was often more communal, kitchens were large and so were the meals. This way of life led to the development of the most important device of this time: the turnspit.

It was to survive as the chief appliance in the kitchen for hundreds of years until, in the late 1700s, the revolutionary idea of roasting meat in an oven was introduced. Not that the spit has entirely died away: even today, the most up-to-date cookers may include a rotary spit, electrically operated.

But in the early days the laborious business of turning the spit for cooking meat evenly led to a number of inventions. One which appeared in the 1500s consisted of ropes and pulleys leading from the spit by the fire up to a wooden drum-shaped cage on the wall. A small dog was put in the cage and as the dog ran, the cage revolved and the spit turned.

Also in the 1500s, the Italian all-rounder Leonardo da Vinci invented a self-turning spit, worked by the heat of the fire going up the chimney. A small turbine wheel was fixed in the chimney with rods and cogwheels connecting it to the spit. As the heat ascended, the wheel turned, fast or slowly according to the strength of the fire.

In time, the invention of clockwork led to a spit-jack, hung above the fire. It was wound by a key and turned the spit slowly as it unwound. The same idea was incorporated later still in the Dutch ovens which lasted well into the present century.

A Dutch oven was a dome-shaped metal box enclosed on three sides and open at the front, where it was placed to face the fire. It had a hook inside for hanging the meat and the clockwork spit-jack above for turning it.

These ideas all depended on one innovation which occurred in the 1500s. That was the movement of the main kitchen fire from the centre of the room to the side, where a chimney could take away the smoke and where it could be enclosed by an inglenook.

This development also made possible an invention which lasted until the development of ovens: the pot-crane. By the 1700s, the pot-crane was quite an elaborate affair of wrought iron with ratchets, bars, hooks and hinges. Fixed to the wall at the side of the fire, it could be swung out into the room, raised or lowered or moved by lever backwards and forwards.

The pot-crane usually incorporated an idleback, an aptly named labour-saving device which allowed the housewife to move a kettle over and away from the fire as required and even tip it for pouring without lifting it off the crane.

A number of gadgets were introduced for inglenook cooking. There were frying-pans and griddles which hung from chains, small trivets, waffle irons, grillers and a tall stand with spikes on which meat was impaled. The downhearth toaster was a long-time favourite. It was a device with a swivel which allowed the toast to be turned when one side was done without removing the bread from the fire.

It must have been the basic design of the inglenook which gave inventors the idea for the first cooking ranges – the next great domestic leap forward.

The idea of the range started with the gradual replacement of wood by coal as a domestic fuel. This substitution led to the invention of braziers or fire-baskets. It was only a matter of bricking up parts of the inglenook to form hobs either side of the firegrate, and something very like the basic kitchen range had been formed.

As early as the 1630s, an Englishman, John Sibthorpe, had patented an oven to be used with coal. But the idea was slow to catch on. Not until the end of the next century was there a flurry of inventive activity which promoted the idea of a grate fire heating an adjacent oven.

One of the earliest inventors whose name we know was the colourful Count Rumford of Bavaria. Born Ben Thompson in Woburn, Massachusetts, in 1753, he

R.M.S.

was a poor boy who became one of Europe's notables. Among other gadets, he invented the first scientific fireplace and flue, a coffee percolator, a water-jacketed roasting pan to cook evenly without burning, and the first hotplates of concentric rings which would adjust to the size of the pan.

But his idea for a cottage-size cooker failed to win favour. It consisted of two cylinders, one inside the other. A fire at the bottom provided the heat, the space between the cylinders was the flue and led up to a chimney on top.

A similar idea of 1770 used flue gases to heat the hotplate in a successful design invented in Devon for making clotted cream.

Around 1780, a British inventor, Thomas Robinson, patented the first true kitchen range. With a cast-iron oven on one side of the firegrate and a water boiler on the other, it included a fuel-saving device to alter the size of the firegrate.

Then in 1802, an English iron-founder, George Bodley, patented his closed-top cooking range. It became the prototype of all the kitchen ranges which continued in service until after World War Two.

The Bodley range's innovation was the design of the flue, which took the heat right round the oven and provided a more even heat. It became a standard piece of kitchen equipment in the 1800s, although another favourite was the American 'portable range', a free-standing model having legs and an iron flue.

This American model could be regarded as the true original of today's free-standing gas and electric

cookers. They have become smaller overall, more streamlined in design, more controllable with thermostats and presetting timers, and much easier to clean; but the basic ideas were laid down in those early days.

In the same year that Bodley brought out his cooking range a German businessman, Frederick Albert Winsor, cooked a meal by gas for the first time. The first gas range was built in the 1840s in the United States, but domestic gas-cookers were not produced in any number until the 1860s.

An early gas appliance which sounds more cumbersome than useful was an 1824 cooker consisting of a gridiron perforated with holes. A frying-pan or saucepan could rest on the top, or the device could be turned on its side to cook meat suspended in front of it.

Plainer and more serviceable gas-cookers became popular in the United States in the 1870s. The first electric cookers appeared in 1890, but it was to be 30 or more years before they became really popular.

The introduction of cheap and easily available electricity in the 1920s and 1930s led to widespread development in electric appliances for the home. It is, however, a sobering thought that most of the electric labour-savers we know today were actually on the market before the end of the 19th century.

They included electric kettles, irons, saucepans, frying-pans, toasters, roasters, hotplates, ovens, coffee-grinders, immersion heaters and fans. Today's refinements include pressure-cookers and infra-red grills which reduce cooking times to minutes, and foodmixers having attachments for slicing, chopping and grinding.

Mass-produced factory foods drove out of the kitchen a number of Victorian gadgets such as ice-cream makers, or freezers, as they were called. These were hand-turned and used ice and salt as cooling agents. Some of them could make ice-cream in five minutes.

Many of today's electric kitchen appliances are based on the ideas of the inventive Victorians whose devices were hand-turned. These included an 1865 dishwasher similar in principle to the modern electric ones, but hand-powered.

The pressure cooker isn't quite so modern as you might think — this model was made in 1890.
Far left: Household items in wrought iron from North American Colonial days — in the foreground whale oil lamps, with a Betty lamp on the left, and in the background an iron crane with cooking utensils suspended from it.

Below left: This early gas cooker was, believe it or not, largely made of wood — which certainly helped insulation, if not safety. *Below:* The early days of electric cooking, in an old-world cottage kitchen.

The old and the new — 1: Invention has taken cooking a long way: *right,* a modern table-top hob, with separate wall-mounted oven and matching refrigerator; *far right,* the first vacuum cleaners required as much 'man-power' as hand-cleaning!

The old and the new — 2: A vacuum cleaner of 60 years ago — certainly an advance on the man-powered machine; *below right,* a modern microwave oven which can cook a joint in literally minutes.

Below: A washing machine of 1884 — strictly hand-operated!

A knife-cleaning machine which was to stand the test of time, lasting well into this century, consisted of a wooden drum with brushes inside which cleaned eight knives a minute as the drum was rotated.

The carpet-sweeper, having brushes which rotated inside a pan as the sweeper was pushed, was the first real advance on the brooms of birch or willow twigs used in the Middle Ages. The sweeper was invented by Melville R Bissell, of Grand Rapids, Michigan, in 1876. The idea was so good that carpet-sweepers are still in use today, often standing side by side in the cleaning cupboard with the vacuum cleaner, one of the most successful inventions for domestic labour-saving.

A vacuum cleaner was invented in the United States as early as 1859, but it was hand-powered. Early vacuum cleaners were worked by bellows, and consisted of a wheeled box and pump handle. One servant was needed to work the bellows, and a second to wield a long hose ending in a small suction fitment.

Like the modern dishwasher, which was used commercially in hotels and restaurants for some years before appearing on the domestic market, the vacuum cleaner was first used for cleaning theatres and other large public buildings. John S Thurman of St Louis, Missouri, invented the first powered vacuum cleaner in 1899, but the familiar carpet-cleaning model for household use did not appear until 1907. Cylinder machines, which were light and versatile, appeared in the 1920s and proved very popular.

Just as house-cleaning is a comparatively modern idea, so is frequent clothes-washing. An early washday appliance was the mangle. Tudor England had a clumsy wooden structure to do the job. A large hand-turned wheel at one side drew a tray weighted with stones over rollers which pressed on the clothes.

The mangles of the 1850s, heavy wooden rollers set in an iron stand and turned by hand through a large iron wheel, continued well into the 20th century. Rubber rollers were the next improvement, first on individual stands and later attached to electrically-driven washing machines.

Today, mangles have largely been superseded by spin-driers, which spin the water out of clothes by centrifugal force, and tumbler-driers which use hot air.

For the household of the 1800s, washing equipment consisted of wooden tubs, a copper boiler – later with a built-in gas heater – a clothes-horse and washing dollies. Dollies had broom-like handles with, instead of a broom-head, a device like a small stool on the end. The dolly was moved up and down and turned to and fro to agitate the washing. A later version, sometimes called a posser, had a copper cone with holes in the side, instead of the stool-like ending.

This simple dolly was the basic washing tool until it was ousted by the modern washing-machine. Indeed, it was incorporated in an electric machine of the 1920s which simply moved a dolly up and down in a wooden tub.

The first washing-machines, however, were hand-powered. A model of the 1840s was simply a wooden box on rockers, not unlike a cradle. Strange as it may seem, it was still selling in 1927.

In 1850, an American, Joel Houghton, invented a cylindrical, hand-operated washing-machine which was based on the principle of agitated water. Then in 1869 appeared the 'gyrator' type, consisting of a wooden tub, cranked handle and rotating blades, which set the pattern for the early electric models. Probably the first electric washer was the Thor, patented in 1907 by Alva J Fisher of the United States.

About the same time, today's automatic washing-machine was foreshadowed by the steam washer – a perforated copper drum enclosed in a copper tank. Gas heated water below it to create steam, but the drum was revolved by hand.

A later improvement allowed the lower part of the drum to pass through the water. An idea which was not pursued was to carry the water up and over the clothes in small containers fixed to the inside of the drum, rather like a conveyor belt. The makers claimed that a fortnight's wash for a family of eight could be done in two hours.

Not until the 1930s did more streamlined electric machines appear, having agitators, power-wringers and better insulation.

World War Two put a stop to washing-machine development, but in the 1950s agitators were

Inventions to help dispel washday blues: *far left*, Thomas Bradford's 'Vowel Y' washing machine and mangle of 1897; *left*, ironing as depicted by an 18th-century artist; *top*, a gas appliance of 1870 for heating flat irons, and *above*, a self-heating iron of 1924.

gradually replaced by impellers, which pushed water through the clothes. The 1960s saw the spread of automatic machines in which pre-set programmes take care of the complete washing cycle from inserting the soiled clothes to the spin-damp stage.

One important invention which made the modern washing process possible was soapless detergent. Soap, itself a detergent – the word means any cleaning substance – was probably invented in Gaul in the first century AD, though legend ascribes it to a mixture of the fat from animals sacrificed on Sapo Hill, Rome, with clay. The black soap of the Middle Ages was a mixture of sand, ashes and linseed oil.

Profit motive

If anyone had asked French chef Nicolas Appert what his motive was in working out his great invention, he would probably have answered '12,000 francs'. That was the prize offered by the French government in 1795 for a method of preserving food for transporting to its armies. Appert experimented for 15 years before arriving at a method of heating food in airtight glass jars — not very different from the home-bottling that housewives do today.

An English inventor, Peter Durand, read Appert's book about his work, published in 1810, and decided to try boxes made of tinplate, which is stronger and lighter than glass. He interested the owners of a London ironworks, Bryan Donkin and John Hall who, scenting a chance to make money from the project, took the idea up. Their investment paid them handsomely. Appert also liked the idea, and used his prize-money to establish the world's first commercial canning factory in 1812.

How long does canned food last? An undamaged can of roast veal that dated back to 1824 was opened in 1938, and the meat was found to be in perfect condition.

Soaps were refined over the centuries, but it was not until the growth of the oil industry that what we know as detergents evolved. A German scientist, Fritz Gunther, made the first soapless detergent in 1916, but it was too harsh for household use. The first domestic detergents were compounded by American chemists working for the Procter and Gamble company in 1933. Chiefly made from petroleum and coal by-products, detergents have the advantage that they will clean in hard or salt water. Non-lather varieties made possible the front-loading automatic machine, and they have continued to be developed over the years.

Irons were among the early inventions for domestic use. The box iron had a compartment for a piece of heated charcoal or iron, or a hot brick. The flat iron was itself heated on the fire. For many years when ruffs and lace were popular, there was the goffering iron, a cylindrical heated bar used with a special stand.

Electric irons were on sale in England as early as 1896. They weighed 14 lb (6kg), so it must have been hard work to use them. The first simple and effective electric iron for household use was not marketed until after World War One.

The principle of using ice to preserve food was exploited by the Chinese 3000 years ago. They had ice-cellars where blocks of ice cut from lakes and ponds in winter were stored between layers of straw. Salting, drying and spicing had to do the preserving work for

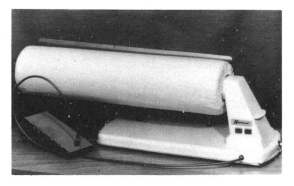

A modern rotary ironer which really takes all the backache out of ironing.

most people in medieval Europe. Ice-cellars were common in countries with hard winters, like parts of Europe and North America, from the 1500s onward.

A simple household icebox of the 1800s was often a wooden cabinet lined with zinc and containing two compartments. One held a block of ice and the other was fitted with movable racks for storing food.

The principle of lowering the temperature of the atmosphere by using ammonia as a refrigerant was discovered by the British scientist Michael Faraday in the 1820s.

In 1834, the Massachusetts inventor Jacob Perkins made the first vapour-compression refrigerator but, as usual, the domestic models lagged a few years behind the commercial product. It was not until the 1870s that the first really practical home refrigerator came on the market. It was a Swedish invention based on Faraday's ammonia principle and worked by gas or oil. About the same time, an American firm brought out a compressor-type model.

Deep-freezers remained a commercial appliance until the late 1930s when the first household models appeared. They were too expensive to be widespread and it was not until the 1950s that simpler, cheaper domestic models came on to the market.

While summer brought its problems in terms of preserving food, winter meant a long hard struggle to keep the home comfortably warm and well-lit. For most people in the cold climates of Europe and North

America, neither of these ambitions was possible until the early 1900s.

The Romans invented a central-heating system, the hypocaust, centuries ahead of anyone else. Based on the circulation of warm air, the hypocaust consisted of channels under the floor and inside the walls of a building, to carry the heat from the furnace. It was fired by charcoal from a stokehole on an outside wall.

By the early 1800s a similar system was installed in many American houses. A large stove in the basement sent hot air through pipes to vents in each room. By the end of the century, the system was based on boilers and hot-water radiators.

For much of history, domestic lamps were feeble, smoky, smelly and dirty. Pottery lamps, burning vegetable oil by means of a simple wick, were used in Egypt around 1300 BC. The Greeks and Romans used bronze and clay lamps of much the same design, with wicks of oakum, linen or papyrus.

The earliest mention of candles was about the first century AD, when they were made of tallow containing a wick. In the Middle Ages, twisted hemp or rushes dipped in tallow were used with lamps of horn or metal, pierced at the sides.

Leonardo da Vinci invented a lamp which helped him work late into the night. A glass cylinder containing olive oil and a wick was placed in a globe filled with water. The light from the burning wick was magnified by the water.

In the 1830s, oil lamps improved with the development of the colza lamp; colza or rape oil, extracted from crushed seeds, was non-odorous – and some of the early oil lamps were very smelly indeed, as were tallow candles. Some 20 years later, kerosene (paraffin) oil lamps were invented.

Gas lighting is much older than is generally supposed. The Chinese used to burn natural gas 3000 years ago to evaporate brine and produce salt, and early fire-worshippers erected temples around natural gas jets. But the first person to manufacture coal gas was a Belgian, Jan Baptista van Helmont, in the early 1600s. The French scientist Antonine Lavoisier planned to light cities by gas, and he invented a gasholder in the 1780s.

The Scots-born engineer William Murdock used gas to light a large factory in 1798, but the use of gas on a large scale for lighting was the work of the German-born Frederick Albert Winsor, who founded the world's first gas company in London in 1813. Early gas lighting was relatively inefficient until 1855, when Robert von Bunsen, a German scientist, developed a burner in which gas and air could be pre-mixed for burning.

One of Bunsen's students, Karl Auer, invented the

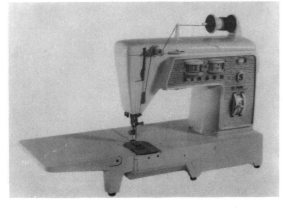

The 'garden engine' was another useful device from the mid-1800s. Consisting of a cistern on a wheeled framework of wrought iron, it had a pump and handle, and was used to water plants and trees.

The match, another household item we take for granted today, was the subject of much experiment in the 1800s. The laborious tinderbox system of flint and steel was the best way of producing a spark until 1805, when the bottle match was invented. A mixture of potash, sugar, gum and sulphur daubed on a splinter of wood, it ignited when dipped in a bottle of sulphuric acid.

Striking matches were invented by an English chemist, John Walker, in the 1820s, and were known as 'Lucifers'. The first phosphorous matches were the invention of a Frenchman, Charles Sauria, in 1830. Safety matches were invented by a Swedish chemist, Gustav Pasch, in 1844, and book matches by an American lawyer, Joshua Pusey, in 1892.

While the introduction of some household appliances was greeted with enthusiasm, others met resistance. Except among the Romans, personal bathing was almost as rare an event as clothes-washing through much of history. It was said of one English queen that 'she taketh a bath twice a year, whether she needeth it or no'.

About the mid-1800s hip baths, filled by servants ferrying jugs of water, were in use. Around the same time, one of the earliest shower-baths was introduced. It looked much like a tall, narrow, curtained tent with a perforated receptacle on top. A line of servants passed forward jugs of water to a colleague on a ladder, who poured the water into the receptacle and over the bather.

Later, all sorts of tin baths appeared, even a portable model having a lid and leather straps. There was one with a gas heater underneath and another, known as a leg bath, which was shaped like a giant boot.

Around 1850, speaking tubes were invented for the home market. These connected various rooms in the house to the servants' quarters. A whistle down the tube was the usual way of calling attention, but John Black, a Glasgow plumber, invented a system of bells to make the signal.

While the servants did the housework, the lady of the house may have read or sewed, and if that exertion proved too much on a warm day, she could cool her

Inventors and manufacturers put a great deal of thought into making apparatus look good by the standards of the day. *Top left*, 'The Triumph' was described in an advertisement of 1891 as 'very powerful'; *above*, an early gas geyser.

Below left: This sinister-looking apparatus was the first zip fastener. Its inventor, American Whitcomb L Judson, called it a 'clasp locker' when he exhibited it at Chicago in 1893. *Below:* A catalogue of 1890 described this object as 'a handsome electro-bronzed duplex lamp;' it sold for less than £1.

gas mantle in 1885. It was made of thread dipped in thorium and cerium nitrate. When first lit, the thread burned away leaving a 'skeleton' of carbonized thorium and serium nitrate which glowed brightly. Similar mantles are still used.

The first electric light was invented by the Cornish scientist, Sir Humphry Davy. It was the arc lamp, which he produced in 1810. But it was too bright for household use, and domestic lighting had to wait for the incandescent lamp, the basis of which is a filament glowing white-hot when current is passed through it. Joseph Swan in England and Thomas Alva Edison in the United States both hit on the idea of using carbon for the filament. Swan patented his lamp in 1878, and Edison registered his patent in 1879. Later, tungsten was found to be the best material for filaments, and this metal is generally used in electric lamps today.

Another invention which probably made life easier for housewives was the sewing-machine, and its commercial use brought down the price of ready-made clothes. Thomas Saint of London patented the first sewing-machine in 1790. It was made of wood and brass, but was not practical.

Another wooden sewing-machine was made by a French tailor, Barthélemy Thimonnier, in 1830. This was more successful – so much so that his workmen saw it as a threat to their livelihood. They destroyed the machines Thimonnier had made, and he was too disheartened to try again.

An American, Walter Hunt, invented another sewing-machine in 1832, but did nothing with it, again because of opposition from tailors and seamstresses who feared for their jobs. Another American, Elias Howe, produced a machine in 1846, and patented it, but failed to find a backer. In 1851 Isaac Singer of New York put a successful sewing-machine on the market, and it was an instant success. Howe was able to claim royalties on these and other machines. The first electric sewing-machine was made in 1889.

Lawnmowers were known in Victorian times, but were operated by hand or pulled by pony or horse.

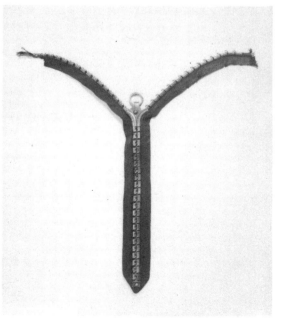

Above, a nightlight provided the heat for this nursery lamp food-warmer of the late 1800s.
Top right: Roman lamps, of a type used in many parts of the Middle East and Europe for thousands of years.
Right: Thomas Edison's electric light, as exhibited at the Crystal Palace, London, in 1882.

face with a pedal *Zephyrion*, a fan on a tall stand which was worked up and down by a foot pedal.

On the other hand, for cold nights, one of the more way-out ideas was a box-shaped bedwarmer consisting of a large openwork iron frame. A charcoal brazier was suspended in the middle of the frame and the whole was put into the bed. A similar idea on a smaller scale was the Victorian 'Instra' pocket-heater incorporating a packet fuel which smouldered in a small perforated box.

One of the many Victorian bedroom gadgets which died a quiet death was the air funnel, a large trumpet-shaped duct through the wall. Its mouth was almost the width of the bed.

Then there was the bizarre alarum bedstead which featured an alarm clock attached to the bedhead. When the bell stopped ringing, the front legs of the bed gently folded and the sleeper, 'without any jerk or the slightest personal danger,' was tipped on to his feet in the centre of the room, where a cold bath could await him.

A more comfortable idea was the 'Dream of Ease'. An elaborate reclining chair with pushbutton adjustability, it incorporated a bookrest and a writing-table.

Clockwork cradle-rockers, shaving machines and a clockwork ensemble of bars, levers, wheels and hinges which served meals were among the ideas patented in the 1890s and never heard of again.

As for the ultimate in gadgetry – the robot maid – a number of prototypes have been invented but have never reached the market. Inventors have been promising for years that the robot servant, who will do all the housework yet cost less than a small car, is just around the corner; but the housewives of the world are still waiting for it.

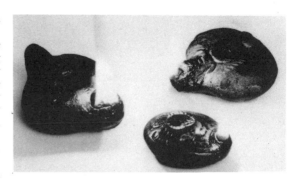

Right: Getting up in comfort has preoccupied people for more than 70 years: this automatic tea-maker was invented back in 1903.

Below: This standard lamp of 1890 was described as 'very good' by its makers; it adjusted in height from 137 cm to 213 cm.

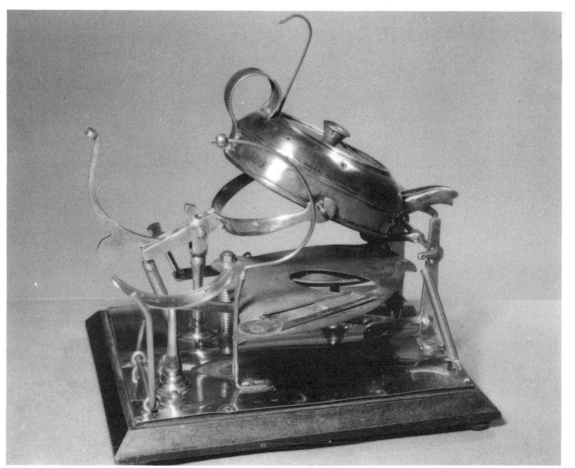

The Penny-Farthing bicycle enjoyed a vogue beginning in 1870. In that year James Starley invented the 'Starley Wheel', in which the rim and hub were connected by tensioned wire spokes. He also settled the relative sizes of the two wheels.

The story of transport is one of dreams fulfilled, obstacles overcome and limitations surpassed. It is also one of ever-increasing speed which, in the 1880s, led to the invention of time zones – until then, people simply set their clocks by the Sun – and which more recently has produced the modern disease known as 'jet fatigue' by confusing Man's built-in biological clock.

Scattered liberally through the story are many weird and wonderful inventions; some bizarre and, though unproductive in themselves, significant for the future, and some just plain odd. Today, nobody would dream, or be able to afford the expense, of designing and building a new car, ship, train or aircraft without first working out the implications on paper, probably with the help of computers. But in the early days of each mode of transport, and particularly during the late 1800s, many enthusiasts and eccentrics, often unencumbered by scientific knowledge or practical experience, did exactly that.

They produced such fantasies as a railway locomotive powered by two horses (called *Cycloped*, it competed against Stephenson's *Rocket* at the famous Rainhill trials in 1829), and a petrol 'tractor' in the shape of a horse, with a pair of driving wheels between the 'horse's' legs. So much for the inventor's supposed overwhelming desire to supersede the horse!

Clement Ader built an improbable looking bat-winged, feather-propellered machine, Eole, which just managed to hop briefly off the ground in 1890. In 1873 the Russians launched a virtually unmanoeuvrable circular ship, *Novgorod*, while an extraordinary British vessel, the *Connector*, which had three hinged sections, took to the water in the 1850s.

Inventions that were to change the world were often treated by press and public as nothing more than toys, and often dangerous ones at that. Richard Trevithick, English pioneer of the railway locomotive, built a successful steam coach in 1803. It carried eight passengers at a steady 12 mph (19 kph). But it frightened horses – and nobody but the inventor could see any use for it. The Wright Brothers' first successful powered flight in 1903 passed virtually unnoticed. Even the London *Daily Mail*, whose proprietor, Lord Northcliffe, was alive to the possibilities of air travel, dismissed the story in 12 lines under the heading 'Balloonless airships'! It took many years to convince the public at large that metal ships could float, and that steam engines could safely propel them across the oceans.

People were often actively hostile to new machines. One English magistrate so hated the new-fangled motor-cars that he took cover behind his garden wall and pelted them with refuse. In Britain motor-cars were restricted to a maximum speed of 4 mph (6 kph) and had to be preceded by a man on foot carrying a red flag, until the repeal of the notorious Red Flag Act in 1896. In the United States the Pennsylvania Farmers' Anti-Motoring Society required a motorist meeting a team of horses to pull off the road and cover his vehicle with a cloth painted to blend with the scenery.

Car, train, ship and air transport have long been part of everyday life and these days, a space-craft launch arouses less interest than a football match. However, it is now our turn to wonder whether mankind will ever accept some of the latest transportation inventions. What, for example, about the automatically controlled automobile on an 'electronic highway', in which the 'driver' sits back and leaves the computers to get on with it? Or the hypersonic airliner travelling at 5000 mph (8000 kph) powered by external combustion and enveloped in a sheet of flame? The giant hovercraft skimming across the Atlantic waves at 100 mph (160 kph) seems almost tame in comparison, as does the 300 mph (480 kph) 'maglev' train which never touches the rails, but is levitated by magnetism.

The latest in commercial jet travel — the Anglo-French supersonic Concorde lands after a test flight.

Railways

The people of ancient Greece discovered over 2000 years ago that pulling a wagon along a smooth 'permanent way' required much less effort than hauling it along a rough road. They made extensive use of rut-ways – narrow grooves cut in the rock in which wagon wheels moved with a minimum of friction. Their 'invention' was not really new. The rock of Malta is criss-crossed with Neolithic 'cart-tracks', deep V-shaped grooves which were apparently used for hauling loads along, though probably on a framework of poles, and not on any wheeled vehicle.

Neolithic Maltese and ancient Greeks had invented the basic principle of the railway, although the modern version, the raised wagonway or railroad, did not appear until nearly 1500 years later.

During the 1500s wagonways were widespread in mines, where use was also made of the railway's second basic principle – that a wagon travelling along a 'guideway' requires less space than a steerable road vehicle. Down the mine this meant that heavily loaded trucks could be safely hauled along narrow tunnels. The early wagonways were made of wood, and the wagon wheels had either an inner or outer flange to keep them on the rails.

In the 1600s the wagonways reached the surface, to link the pithead to the nearest river, and in the 1700s the not very permanent wooden 'way' was replaced by metal. Workmen called platelayers fixed iron plateways to the wooden rails. These plateways were flanged, so that smooth-wheeled wagons able to run on road or rail could be used. But the problems of making points with flanged rails led to a return, early in the railway era, to plain rails and flanged wheels.

The early coal wagons were horsepowered, as was the world's first regular passenger service, inaugurated in 1807 on the Swansea and Mumbles Railway in south Wales. After a few weeks of service, the line experimented with sail power – the 4½ mile (7 km)

Richard Trevithick's 'Catch-me-who-can', which ran for several months in London's Euston Square as a public attraction.

journey took 45 minutes – but soon reverted to the horse, and did not use steam locomotives until 1877. In 1830, both the Baltimore and Ohio and the South Carolina railroads tried sail power with – when the wind was in the right direction – great success, and in places sailing trucks were used by maintenance gangs for many years. But the ultimate success of the railways depended on steam power.

The world's first railway locomotive was built by a Cornish engineer, Richard Trevithick, in 1804. Its single-cylinder engine turned a massive flywheel through a 'trombone slide' linkage, while a system of toothed wheels linked the flywheel to the driving wheels. Called *Penn-y-darran*, it was a triumphant success. But it was too heavy for the existing track and was abandoned.

Many engineers believed that smooth metal driving wheels would fail to grip smooth metal rails, and the first commercially successful locomotive, built by another Englishman, John Blenkinsop, in 1812, was driven along by a toothed wheel which engaged in a rack on the rail. Only a year later a Tyneside engineer, William Hedley, built *Puffing Billy*, a friction-drive locomotive which was used to haul coal trucks to the

Second-class travel in 1833 on the Liverpool and Manchester Railway provided only minimum protection for the passengers.

Above: The most famous of all pioneer locomotives — George Stephenson's 'Rocket', which won the Rainhill Trials in 1829.
Right: William Hedley's 'Puffing Billy', built in 1813 and used for hauling coal trucks in the north of England, proved that metal wheels would grip on metal rails.
Top right: 'Hibernia', an early locomotive on the Dublin and Kingstown Railway in Ireland.

Locomotives were not the only means of operation in the pioneer days of railways. The London-to-Blackwall line had its trains coupled to cables, which hauled the carriages along.

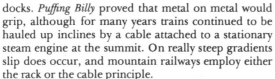

docks. *Puffing Billy* proved that metal on metal would grip, although for many years trains continued to be hauled up inclines by a cable attached to a stationary steam engine at the summit. On really steep gradients slip does occur, and mountain railways employ either the rack or the cable principle.

The world's first steam-powered public railway, the

Stockton and Darlington, in north-east England, opened in 1825. Its first locomotive was George Stephenson's *Locomotion*, but this engine and others like it proved slow, unreliable and expensive. Stephenson, a Newcastle engineer, accordingly designed the *Rocket*. Tough, reliable, and with an efficient multi-tube boiler, it hauled trains on the world's first fully-fledged public railway, the Liverpool and Manchester. Stephenson's *Rocket* and his *Planet* (1830) were the true ancestors of most later steam locomotives. Radically different early designs, such as the vertical boilers of the British-built *Novelty* (1829) and the pioneer New York-built *Best Friend* of Charleston (1830), were soon abandoned – as was the French engineer Marc Seguin's curious invention of 1829.

France's first steam locomotive, Seguin's machine had a huge pair of bellows on the tender to improve the draught in the furnace. Unfortunately the bellows were driven by the wheels, so were least effective when building up steam or climbing a gradient – the two occasions when they were most needed.

The year 1832 saw the invention of the bogie

Signal success

The first railways needed no signals but, as traffic expanded, a system of control had to be introduced. According to railway tradition, the first signal was invented by one of the stationmasters on England's Stockton and Darlington Railway, in the late 1820s. If he put a lighted candle in the window of his house, the drivers stopped at the station. If there was no light, they went straight through!

Another pioneer line, the British Great Western, used a ball hung from an arm on the end of a post. If the ball was visible, the line was clear. At night, a lantern took the place of the ball. The familiar semaphore signal, still in use in some parts of the world, was introduced in 1842.

The modern light system of railway signalling was invented in 1913, but it was not introduced until 1921, when a system of light signals was installed on the Liverpool Overhead Railway in England.

Electric power was first used to operate signals and points in 1884, when the American inventor George Westinghouse devised suitable apparatus and installed it on the Philadelphia and Reading Railroad in the United States. Westinghouse invented several devices for the automatic control of signals and points.

Left: A sloping train for a sloping track — a rack railway in the Swiss Alps.
Right: Tilting for speed — British Rail's Advanced Passenger Train demonstrating its body-tilting capabilities during trials. It is designed to travel at 250 kph on existing track.

locomotive *Experiment* in the United States. The wheels of the swivelling bogie follow curved or uneven track much better than fixed wheels, and permit greater speeds.

Already the standard gauge of 4 feet $8\frac{1}{2}$ inches (1.435m) had been widely adopted in Europe and America. This curious dimension is said to derive from the 'gauge' of Roman chariot wheels. At one time 23 different railroad gauges were used in the United States, and Australia is still plagued with gauge differences. The engineer-designer Isambard Kingdom Brunel decided on a 7 foot (2.1m) gauge for Britain's Great Western Railway, to allow greater

A typical American locomotive of 1875. Engines such as this hauled trains across North America in the Civil War period and for many years afterwards.

Below, left: The world's fastest regular train service, the 'Bullet', the Tokyo-Osaka super express which covers 515 km at speeds up to 200 kph. The journey takes just three hours.
Below, right: A typical modern diesel locomotive of the Canadian Pacific Railways.

speed and stability. The GWR's trains were for many years the fastest and most comfortable of all.

An unusual experimental express locomotive, the *Hurricane*, was built for the GWR, in which engine and boiler were separated so that each could be better designed. But, freed from the weight of the boiler, the huge driving wheels had no adhesion.

Another of Brunel's inventions was the atmospheric railway, in which the entire train, or a piston to which the train was connected, was drawn along a tube by suction. Ahead of the piston, air was sucked out of the tube so that atmospheric pressure behind pushed the train forward, without noise or smoke. The idea worked splendidly, until rats gnawed holes in the vital leather flaps that sealed the tube – and with no seal, there was no suction.

Perhaps the most bizarre of all strange trains were those designed by Charles Lartigue to run along an A-shaped trestle-like track. Some ran on the Listowell and Ballybunion line in Ireland. Locomotives and rolling stock were double, hanging like panniers on either side of the track. Naturally the loads on each side had to balance – and there is, of course, a story of a farmer who wanted to send one cow. He had to balance it with two calves, which returned one on either side.

With advances such as improved valves, superheated steam and compound engines, the steam locomotive grew ever more powerful and efficient. It reached a peak in 1938, when the British streamlined *Mallard* achieved an all-time speed record for a steam-driven train of 126 mph (203 kph).

Among the most wonderful steam locomotives were the articulated giants designed to haul heavy loads on light and curving track. The most powerful ever, an American triplex, Mallet, had a total of 30 wheels. The Mallet type was named after its inventor, Anatole Mallet, a Swiss engineer. Another type of articulated locomotive, the Garratt, had a massive boiler slung between two engines. Garratts were extensively used in Africa and the Soviet Union in the 1930s.

The first successful electric locomotive was the brainchild of a German inventor, Werner von Siemens, who made many innovations in electrical engineering. Today the Japanese Tokaido-line electric 'bullet' train averages 112 mph (180 kph). Diesel locomotives followed swiftly on the invention of the diesel engine itself (see page 107), and a pioneer model was built in Germany in 1912. The first diesel-electric rail-car was designed in Sweden in 1913.

Modern inventions are aimed at improving speed and performance. Britain's Advanced Passenger Train (APT) can do 155 mph (250 kph) on existing tracks with the help of an automatic tilting mechanism to ensure comfort and safety on bends.

For the future, there are experimental hovertrains which skim just clear of a concrete guide-rail. The French gas turbine airscrew-propelled *Aerotrain*, to date the world's fastest train at 233 mph (375 kph), is one example. Then there are proposals for overhead monorails, with the train perched on top of, or slung beneath the single rail.

The most wonderful invention of all is a combination of magnetic levitation, in which the train is held off the rails by opposed magnets, and linear induction, in which the train is propelled by magnetic forces. There are no moving parts, apart from the actual train and, since the train does not touch the rails, there is no friction. Given a perfectly straight track, speeds of over 300 mph (480 kph) are expected. The Japanese have built a model version, and hope to have a 'maglev' train in service between Tokyo and Osaka in the early 1980s.

Road

A horse-borne litter of a type much used for road travel in Europe and Asia during the Middle Ages.

Mankind's first land vehicle was probably the sledge, a useful device but very heavy going except where there was snow or ice to reduce friction. Then, possibly about 5000 years ago, the wheel was invented. The essential feature of virtually all land transport, it enabled Man and animals to pull heavier loads with less effort.

Eventually a pivoted axle was developed to facilitate steering, a suspension system was added for the comfort of passengers, and harnesses were improved to enable a draught animal to exert its full strength without strangling itself.

There followed a delay of several centuries before roads reached a sufficiently advanced stage for there to be anything more than a limited use for wheeled vehicles, and a further delay before mechanical propulsion appeared on the scene.

Meanwhile, the inventors were busy. In 1599 the Dutch mathematician Simon Stevin built a two-masted sailing wagon. The idea had been tried before, and would be again, but except on ice it has never had much success. However, a similar vehicle, favoured by the wind, sped through the streets of Paris one day in 1834, and another, drawn this time by kites, and called the *Flying Chariot*, could average 15-20 mph (24-32 kph).

Throughout the 1600s and 1700s people experimented with man-powered carriages driven by levers or pedals. They realised only later that horses were stronger and cheaper. Then there was a 'car' powered by a windmill which drove mechanical legs that were supposed to walk it along, and there were clockwork carriages. This pollution-free notion has been suggested again in recent years. But calculations show that a clockwork motor strong enough to propel a car carrying four people at 10 mph for 20 miles (16 kph for 32 km) would need a five-ton mainspring – and eight hours' work to wind it up!

With the development of steam power and of well-made roads in the late 1700s, the possibility of and potential for mechanically powered road vehicles had at last arrived. The first in action was the French engineer Nicolas Cugnot's massive steam tractor of 1769. Designed to haul guns, and with boiler and engine mounted on the single front wheel, it was cumbersome, very hard to steer, and it had to pause every 15 minutes to build up steam. But for the first time the up-and-down reciprocating action of the steam engine had been successfully converted into the rotary action needed to turn a wheel.

Cugnot's vehicle was followed by Richard Trevithick's steam carriages of 1801 and 1803, the world's first practical powered road vehicles, and in the 1820s and 1830s by those of two other Britons, Sir Goldsworthy Gurney and Walter Hancock. The Duke of Wellington said 'It is scarcely possible to calculate the benefit we shall derive from this invention'.

For a while, gaily-painted steam coaches operated successful passenger services in parts of southern England, but by 1840 they had vanished. Opposition from horse-drawn carriage operators, and competition from the newly invented railways, forced them off the road. A few wealthy individuals built or bought scaled-down versions, the first steam cars, but

The Pedestrian Hobby-horse was first seen in Paris in 1816. This version, the Draisine, was invented two years later and imported from France to England by Denis Johnson of London, here seen riding it.

A Chinese palanquin of the 1800s represents a type of transport used when roads were poor and travel was much slower than it is today.

Above: Kirkpatrick Macmillan's lever-driven bicycle was invented in 1839.
Right: An invention that got away — Robert William Thomson's pneumatic tyre of 1845, which was forgotten until John Boyd Dunlop re-invented it in 1888.

An early attempt at a steam motor-cycle: the studied nonchalance of the rider may well have been bravado.

Below, left: Nicolas Cugnot, a French engineer, built this steam tractor, designed to haul cannon, in 1769. It is regarded as the first true motor vehicle.
Below, right: 'The boneshaker' designed by the French firm of Michaux in 1861.

and popularity. This was the bicycle. In its earliest form, the *célérifère*, the bicycle consisted simply of two cart-type wheels joined by a wooden bar on which the rider sat. Pictures of machines of this type are known from ancient Babylon, Egypt and Rome. There was no provision for steering, no suspension, and no pedals. The rider 'walked' the thing along. The *célérifère* was revived in 1791 in France by the Comte de Sivrae, and was succeeded in 1817 by the *Draisine*. Invented by Karl von Drais, the *Draisine* was propelled in the same way, but could be steered.

A hand-cranked tricycle was invented in 1839, and the first mechnically-driven bicycle was made by a Scottish blacksmith, Kirkpatrick Macmillan, in the same year. It was propelled by moving the feet back and forth to work long levers attached to cranks on the axle of the back wheel. At last Man could propel himself faster than was possible on foot; but Macmillan's machine failed to arouse much interest.

In 1861 a Parisian coachmaker, Pierre Michaux, was asked to repair a *Draisine*. His son casually suggested that cranks and pedals on the front wheel would improve the machine, and the boneshaker or velocipede was born. It promptly went into production, and equally promptly was described by *The Times* of London as 'a new terror in the streets'.

The only way to increase the velocipede's potential for speed was to increase the size of the front wheel, a process that soon led to the so-called 'penny-farthing', or high bicycle. It derived its nickname of 'penny-farthing' from the resemblance of its two dissimilar wheels to two British copper coins of the day. Although unstable, and positively dangerous when braking, it enjoyed an enthusiastic vogue. More cautious riders had to wait for the safety bicycle of 1874, invented by the British engineer H. J. Lawson. His design has remained virtually unchanged to this day, though a machine with small wheels, the *Moulton*, a revolutionary design, was introduced in 1962.

The safety bicycle became a practical proposition because of another invention: the pneumatic tyre. The first such tyre was the invention of the Briton Robert William Thomson in 1845. But he did nothing with his idea, and it had to wait for 50 years until it was revived in 1888 by John Boyd Dunlop, a Scottish veterinary surgeon. The modern tyre is the result of a series of further inventions, including wire edges, invented by A T Brown and G F Stillman of New York State in 1892, and cord reinforcement, the invention of John Fullerton Palmer.

There is one link between the motor-car and the bicycle: the differential gear. This device, which enables the two rear wheels of a motor-car to rotate at different speeds while still applying drive, was invented

for the most part the constructors turned their attention to the development of traction engines.

These steam locomotives of the road were not supplanted in heavy road haulage until the 1920s. Most of them displayed a glorious array of gleaming rods and levers. One extraordinary version had wood-covered driving wheels 15 feet (4.6m) wide.

Meanwhile, the invention that was to create a market for the motor-car, by giving the ordinary man a taste of the fun to be had from a personal means of transport, experienced a rapid increase in efficiency

for the bicycle's three-wheeled cousin, the tricycle. It was the work of the Sussex-born engineer James Starley in 1877. Starley is often called 'the father of the cycle industry' because of his work in producing machines for the mass market.

The story of the motor-car – as opposed to the steam-car – began in 1805, when the Swiss engineer Isaac de Rivaz replaced the steam engine in his horseless carriage with an explosion motor. Exploding gas forced the piston up, gravity pulled it down again – and the 'car' moved slowly across de Rivaz's workshop. It was ingenious, but with hand-operated ignition and valves, and a single-acting cylinder (there was no second explosion to force the piston down again), it was not a practical proposition.

The world's first practical internal combustion engine cars were those built by the German engineers Karl Benz in 1885 and Gottlieb Daimler in 1886. Benz's car, a light three-wheeler, was designed as an entirely new kind of vehicle, but Daimler simply removed the shafts from a four-wheeled carriage and fitted an engine. Both vehicles aroused considerable interest, and soon Benz, Daimler and others had cars in production. The automobile age had begun.

Many early cars had cart or bicycle-type wheels and folding hoods, or no hoods at all. Steering was by tiller, drive by belts, and starting often involved lighting a hot tube ignition burner and trimming and adjusting a wick in the carburettor.

The first 'modern' motor car, with engine at the front under a bonnet, a gearbox, a foot-controlled clutch and rear-wheel drive, was the Panhard-Levassor of 1891, the work of two French carriage-builders, René Panhard and Emile Levassor. Another French engineer, Louis Renault, built the first fully enclosed car in 1898, and followed it eight years later with an eight-feet-high (2.4m) model specially

designed for a gentleman who liked to keep his top hat on. Renault also invented the drive shaft, replacing the chains used to take power to the rear wheels. The first mass-produced car, the American Oldsmobile *Curved Dash* buggy, arrived in 1903, mass-production being the brain-child of Ransom E Olds, an American car manufacturer.

The early years of the motor-car produced many bizarre designs. The *vis-a-vis* was notable for its friendliness; front and rear seats were 'face-to-face', giving the driver a splendid view of his passengers and doubtful forward vision. At the other end of the scale was the unusual De Dion tricycle and trailer combination, in which the passenger followed behind in splendid isolation.

With the *vis-à-vis*, one could distinguish the driver – he was the one with his hands at the controls. But, to

Top left: Church's steam carriage, operating in the 1830s, plied between London and Birmingham.
Top right: Goldsworthy Gurney's steam carriage, built in 1829, was a record-breaker in its day: it did the journey between London and Bath at an average speed of 24 kph.
Above: Henry Ford's first car, the Quadricycle, completed in 1896 and still preserved at Dearborn, Michigan.

Blind Jack

All the inventions in road transport which have led to the present dominance of the motor-car would have been useless without the pioneers of good roads in the 1700s and 1800s.

Among them was 'Blind Jack of Knaresborough', a tall, rangy man who lost his sight in 1723 at the age of six. His handicap did not stop him being in turn a soldier – he fought at Culloden — a smuggler and a stagecoach driver. And in 1765 he pioneered the construction of properly designed roads in northern England. His real name was John Metcalf, and he was born at Knaresborough, Yorkshire.

The historic Model T Ford car: first produced in 1908, more than 15,000,000 of these cars were made by the time production ended in 1927.

Right: A Panhard of 1894, an early model of René Panhard, one of France's pioneers of motor-car invention.
Far right: An NSU Ro80, a car of the 1970s powered by the revolutionary Wankel rotary engine.

With the ever-soaring cost of oil, inventors are working on the problems of efficient electric-powered cars. *Above,* the highly-compact Ford Comuta car, and *right,* the chassis showing the four powerful batteries which drive the vehicle.

jump forward a few years, the matter was more difficult with the French *Bédélia* cyclecar of 1911. This extraordinary yet very popular vehicle had two seats, one behind the other. The 'passenger' in the back controlled the steering, the one in front changed gear – using a long piece of wood to shift the driving belt from one pulley to another. There was no driver, as such. Powered by motor-cycle engines, and with the minimum of bodywork and refinement, cyclecars enjoyed a brief vogue until replaced by vastly superior big cars in miniature, such as France's Bébé Peugeot in 1912, and Britain's Baby Austin in 1922.

The ultimate in between-the-wars luxury was the Italian Bugatti *Royale*, the largest production-car ever. Its 13-litre engine was so powerful that a gearbox was hardly necessary. There was a first gear, for starting; a normal gear, for speeds from 3 to 70 mph (5-112 kph); and an overdrive giving a maximum speed of over 120 mph (190 kph). Only six of these cars were built, and the radiator mascot was appropriately a white elephant.

Engines and other parts were steadily improved, and refinements such as heaters and automatic transmissions were added. One theoretically helpful invention was the *Pasolite*, of 1922. At that time it was customary when passing another car at night to switch off the headlights. With a *Pasolite* fitted, the driver turned a switch which extinguished the headlights and turned on a powerful side-pointing light, thus lighting up the road for the other car – a neat idea, had every car been equipped with the device and used it.

Revolutionary ideas were plentiful, particularly in connection with the number of wheels desirable on a car. Some thought four too many; others, such as the builders of the 1911 American *Octoauto*, wanted more. The *Octoauto* had eight wheels, six of which were by an ingenious mechanism used for steering. Baron von Eckhardstein was content with six wheels on his ostentatious de Dietrich of 1903. The car was luxuriously furnished, and contained a kitchen for the preparation of meals en route.

The Briggs and Stratton *Flyer* of 1920 had five wheels. Advertised as 'one of the most popular contrivances ever brought out for Young America', it resembled a child's go-kart with a fifth wheel added at the back. This extra wheel was the driving wheel. Mounted beside it was a lawn-mower engine giving a speed of some 25 mph (40 kph). Declutching was done with a lever which lifted the driving wheel off the ground.

The two-wheeled gyroscope-balanced *Gyrocar*, built by Wolseley for the Russian Count Schilowsky in 1912, was a successful if pointless novelty. Much more interesting was the concept of Gabriel Voisin, the French aircraft-designer, of a car with four wheels in diamond formation. Single wheels at front and rear were both to be used for steering, giving fantastic manoeuvrability, while driving was through the middle wheels. This idea has been reconsidered in recent years.

Many of the weirdest and most wonderful cars have since the early days been those designed for speed. They range from the streamlined electric *La Jamais Contente* of 1899 (65.79 mph; 105.88 kph) and the Stanley steamer *Wogglebug* of 1907 (150 mph; 240 kph) through aero-engined giants such as the Fiat *Mephistopheles* of 1924 (146 mph; 235 kph) all the way up to the rocket-powered *Blue Flame* which has held the world land speed record since 1970 (631.37 mph; 1016 kph), and back down to the diminutive slingshot dragsters which accelerate to speeds of well over 200 mph (320 kph) in just a quarter of a mile.

Conventional racing cars have over the years pioneered many advances later adopted on normal cars, but recently they have become so specialized that they possess little resemblance or relevance to the family car. Held to the ground by aerofoils (in effect, miniature upside-down aircraft wings which provide downwards thrust instead of lift), they can corner and brake amazingly fast. In 1970 the 'hoover car' made a brief appearance on the racetrack. It worked like a hovercraft in reverse, sucking up air to hold itself down. The idea worked well, but was banned by the racing authorities.

The high speed of the electric- and steam-powered record-breakers reflects the early popularity of both types of propulsion. Leon Serpollet's invention in 1897 of a steam car with a flash boiler did away with bulky boilers, roaring fireboxes and the prolonged delay while steam was built up, and for many years steamers were smoother, quieter, easier to start and simpler to drive than petrol cars. They are, however, complex to build and to service.

Electric cars are also very easy to drive, have excellent acceleration, and they generate little noise and no fumes. But present batteries limit their range to at most 60 miles (95 km). Although the petrol-engined car had won the day by the early 1920s, both steam and electric vehicles are currently being seriously reconsidered.

OI–D

Water

Dug-out canoes, hollowed by fire and hard work from logs, provided one of Man's earliest forms of water transport, and one that is still used in some parts of the world.

In ships like these, those intrepid seafarers the Vikings made their incredible ocean voyages during the Dark Ages. The drawing, *left*, shows a typical merchant ship used in the Baltic Sea, while *above* is an actual Viking longboat, skilfully preserved in Oslo.

Man first took to travelling on water crouched on a log of wood and using his hands to push himself along. Later he found he could move more easily by using a piece of wood. He next fashioned the piece of wood into a paddle – which was almost certainly the first invention in the history of sea transport.

With the help of a paddle, Man could travel faster. The paddle also helped him to steer or control the floating log more skilfully. But a log is a very uncomfortable and unsteady form of vessel. Early Man discovered that a flat piece of wood was much more stable in the water. A flat surface also provided more space, so that two persons and perhaps even their weapons and tools could be carried.

This discovery led Man to invent the dugout canoe. He used fire or primitive tools to hollow out the log, shaped the lower part to make it more stable and produced the first true form of boat.

All this happened tens of thousands of years ago. Historians know little of the early beginnings of ships and travel by sea, because it is perhaps the most ancient form of communication. However, they believe that the next major development in sea transport took place around the 6000s BC. By that time Man had learned that paddling is tiring and made long journeys exhausting. So he learned how to exploit the power of the winds by inventing the sail.

The first sailing ships were made by the Egyptians. They were built of reeds, the most plentiful shipbuilding material to be found in ancient Egypt.

These early ships sailed on the River Nile and ventured out into the Mediterranean Sea.

A ship of this kind had one mast which carried a single square sail. It also carried a crew of oarsmen to row the ship along when there was not enough wind to make the sail work.

At about the same time, the people of the eastern Mediterranean island of Crete were also building ships with sails. These ships were made of wood, of which there was a plentiful supply from the Cretan forests.

By 1000 BC the most successful seamen in the Mediterranean were the Phoenicians, a trading people who lived in the eastern regions of that sea. The Phoenician ships were made of timber and carried one sail. But they also had two decks, each carrying more oarsmen than the Egyptian ships. These ships were known as biremes, and they sailed on long trade voyages. The Phoenicians used to go beyond the Straits of Gibraltar to the British Isles. There they would exchange cloth, dyes and pottery for tin.

In the 400s BC the Greeks and the Romans invented triremes, ships with three decks of oarsmen. The triremes were about 140 feet (40m) long and had 20 oarsmen on each side. They were used in sea battles, and special raised structures, known as castles, were added to them. The castles provided archers with

vantage points from which to shoot their arrows at the enemy.

The castle still exists on modern ships. It is known as the forecastle or fo'c'sle, and is located at the bow of the ship.

In the Baltic Sea and in the northern Atlantic Ocean the Vikings sailed longships equipped with a single sail and a crew of oarsmen. Historians believe these Scandinavian seafarers crossed the Atlantic Ocean to America in AD 1000, almost 500 years before Christopher Columbus. The Viking longships were about 80 feet (25m) long.

The large square sail needed a fairly strong wind to make it work successfully. If the wind was weak, the boat could be moved only by rowing, and the boat could not travel so fast and so far as it did with a good wind pushing its sail.

This Mediterranean carrack of 1490 is similar in size to the 'Santa Maria', Christopher Columbus's flagship on his epic voyage to the Americas in 1492. This drawing shows clearly the fore and aft castles invented for the protection of fighting men.

The next invention in sea transport was the lateen sail. This was a small triangular sail fixed to a yard or tapering pole, placed at an angle of 45 degrees to the mainmast. The lateen sail could work quite well with only a gentle wind, and the Arabs were the first to use it in about AD 1000. Able to take more advantage of the wind, they sailed on long voyages to India and the distant islands of the East, which were rich in valuable spices.

By the 1500s the system of sails on ships had developed to include both square and lateen sails. Ships had three masts: the foremast near the bow, the mainmast in the middle of the ship, and the mizzenmast near the stern or the rear tf the ship.

These sails were controlled and adjusted by the rigging, a system of ropes. The ropes could be slackened or tightened to catch the wind, or to trim the sail for the ship to sail in whatever direction was required.

Advanced forms of rigging were first invented in the

1300s. By the 1700s, rigging was so perfected that a ship could sail straight ahead with a strong wind blowing from the side or from almost straight ahead.

The basic design and structure of ships changed little from the 1500s to the 1700s. Ships became larger and transported more goods and passengers than earlier designs. They were used more and more in wars and carried heavy cannons and large bodies of troops.

World trade began to expand from the 1750s. The need grew for very fast ships to transport goods such as tea, spices and cotton from distant parts of the world to Europe. This led to the development of clipper ships like the *Cutty Sark* and the *Flying Cloud.* These ships had many sails of different shapes and sizes. They reached speeds of up to 20 knots, but could carry only comparatively small cargoes.

The days of the clippers marked the end of the age of sail. The finish came when scientists discovered how to harness the power of steam by inventing the steam engine.

The first mechanically successful steam engine for a ship was invented in 1787 by John Fitch, an American engineer. The engine worked six paddles and produced a speed of 5 mph (8kph). In 1807 another American engineer, Robert Fulton, designed the *Clermont,* another paddle steamship which also travelled at 5 mph.

Fulton's engine, however, was more efficient than Fitch's. It used a circular motion, to drive the engine.

Transport

Light work

Safety at sea has been a prime consideration ever since Man first sailed out of sight of land, and the lighthouse has been one of the most important aids to safe navigation. It was probably invented by the Greeks of Alexandria in Egypt, who built a lighthouse, called the Pharos, about 280 BC. It was one of the Seven Wonders of the World.

The light of the Pharos was a wood fire, and even in the 1700s clusters of candles were the best form of lighthouse beacon that could be managed. At last in 1784 a Swiss engineer, Aimé Argand, of Geneva – a city remote from the sea – came up with an invention that revolutionized lighthouses. It was the Argand burner for oil lamps, using a cylindrical wick inside a glass chimney. Argand burners remained in lighthouse use for more than a century.

Fitch's engine worked with a system of chains and ratchet wheels. The crankshaft is still a basic part of modern steam engines. The *Clermont* operated a highly successful passenger service between New York and Albany in the United States.

Paddle steamers were a major advance, but they suffered from several disadvantages. In heavy seas they wallowed about in the waves. The paddles would rise out of the water and rotate without providing any moving power. And because the paddles were exposed, they could be easily damaged.

To overcome these problems, the Swedish-American engineer John Ericsson invented the first screw propeller in 1839. The screw propeller always stayed in the water and kept working whether the sea was rough or calm. It was also not so exposed to damage as the paddle.

In 1845 the *Great Britain,* a steamship designed by the British engineer Isambard Kingdom Brunel, became the first screw-propeller ship to cross the Atlantic.

The next development in sea transport also came in the early 1800s: the iron ships. These ships were stronger than the wooden ones. Wood was becoming scarce and expensive, and it had to be used in thick sections. Iron could be made up into thin sheets, which left more space for cargoes. Iron ships also stood up better to buffeting and collisions.

The first iron ship of which a complete record exists was the *Vulcan.* The invention of Thomas Watson, a Scottish engineer, it went into service in 1818 as a coastal coal-carrier and worked for more than 60 years.

The first practical seagoing iron ship was the *Aaron Manby,* built in Birmingham in 1821. It sailed to France shortly after its completion. Brunel's *Great Britain* was also an iron ship.

The iron ships needed more powerful engines to drive them than the wooden forerunners. To meet these demands, British engineers invented the compound engine in 1854. This engine produced extra power from the steam by passing it along a series of cylinders.

In 1894 the British engineer Charles Parsons invented the turbine engine (see page 105) and fitted it into the ocean liner *Victorian.* The turbine with its rapid spinning power soon increased the speeds at which ships travelled.

The invention of the diesel engine, first developed by the German engineer Rudolf Diesel in the late

1800s, led to the invention of the first marine diesel in 1911. The engine, which works on the principle of internal combustion, was fitted to the tanker *Vulcanus* and used oil as fuel. From 1918 oil rapidly replaced coal as the main fuel for sea transport.

The most recent development in ship engines is the use of nuclear power, provided by a nuclear reactor in which the energy released from the splitting of uranium atoms is used in a controlled manner. Nuclear engines need only very small quantities of fuel and can travel for many months without refuelling. However, they have to be specially protected because the radiation which nuclear fuel gives out is extremely dangerous to life.

The United States launched the first nuclear-powered vessel in 1954. It was the *Nautilus,* a submarine which made the first North Pole crossing under the ice. In 1959 the Soviet Union launched the *Lenin,* the world's first nuclear-powered icebreaker and in 1962 the United States produced the *Savannah,* the world's first nuclear-powered merchant ship.

The invention of the submarine dates from about 1620. In that year the Dutch scientist Cornelius van Drebbel demonstrated a leather-covered rowing boat that could submerge.

In 1775 an American, David Bushnell, built a one-man, hand-powered submarine which he called the

'Turbinia' was the first ship to be driven by steam turbine, invented by the British engineer Charles Parsons in 1894. Soon after, he fitted turbines into the ocean liner 'Victorian'.

The 'Hunley' was one of the great pioneer submarines. Made from an iron boiler, she was used by the Confederate Navy during the American Civil War of 1861-65. Her eight-man crew drove the boat along while submerged by hand-cranking the screw, as shown in the diagram. 'Hunley' was destroyed while sinking the Union warship 'Housatonic' in 1864.

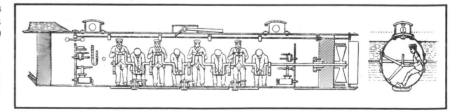

52

Even the conventional speed boat is the subject of work by inventors. *Right,* this boat has a normal hull, but is powered by two smooth-running Wankel rotary engines. *Far right,* the Sea Knife has an entirely new type of hull, developed from a blend of hydrofoil, aircraft and planing hull technologies. It is exceptionally stable and can ride effortlessly through rough water.

Turtle It submerged by letting water into ballast tanks like modern submarines. The first successful electrically driven submarine was invented by Claude-Desiré Goubet of France in the 1880s. In 1898 the French launched the *Gustave Zédé,* 160 feet (50m) long and the largest submarine built up to that time.

When Man first embarked on sea voyages, he had no instruments to show where he was on the ocean or in what direction he was travelling. He plotted his position and course from the stars or the Sun. Sometimes he would determine direction from a steady wind, which he knew came from a certain point such as the west or the south.

The first invention that helped Man to navigate was the magnetic compass. Historians believe the Chinese were the first people to use the compass in about AD 1100. The first compasses employed a needle magnetized by rubbing against a lodestone, which is itself magnetic. The needle was placed on cork or some other buoyant material and floated in a bowl of water; the needle point swung to the magnetic north.

The first reference to the magnetic compass in Europe was by the English scholar Alexander Neckham. In 1187 he wrote of a needle which showed sailors where the North or Pole Star lay when the skies were cloudy. The invention of the marked compass, showing 32 direction points on a circular card, came in the 1300s.

On iron ships the magnetic compass was not reliable, and it was also inaccurate. In the early 1900s scientists invented the gyrocompass, which could work accurately on metal vessels and pointed to the true north. This compass works on the principle of the gyroscope, a rapidly spinning top which is influenced by the rotation of the Earth, and is motivated by an electric motor. The axis on which it spins points always to the north, no matter how much the ship may be rolling about.

The US Navy submarine 'Nautilus' is one of the modern generation of nuclear-powered undersea vessels. On August 1-5, 1958, the 'Nautilus' made the first under-the-ice crossing of the North Pole, sailing from Point Barrow, Alaska, to the Greenland Sea.

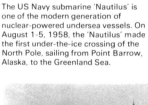

Another major invention in navigation was radar. This invention uses radio waves to spot objects at night, in bad weather or in the far distance. The system works somewhat in the way a bat sends out high-pitched squeaks when it is flying. When the squeaks hit an object, they send back an echo which the bat picks up in its ears; from the echo the bat can tell how far away the object is and how large.

There was no single inventor of radar. Groups of scientists in several countries were working on this project at about the same time. The earliest system resembling modern radar was devised in 1904 by the German scientist Christian Hülsmeyer. His system had a range of only one mile (1.6km), and although it worked it was never adopted. The full use of radar was not made until World War Two. See also page 24.

Sonar, the method used to detect objects under the sea, was invented earlier than radar. Instead of radio waves, this system uses acoustic or sound waves, which bounce off submerged objects and return to the pursuing ship or submarine, where they are picked up and amplified through a loudspeaker.

The first passive sonar system was used in 1916 to pick up the sound of submarine engines. The first active system, used two years later, transmitted sound and bounced it back from the object to be detected. Sonar is also used to locate shoals of fish.

The most recent invention in sea transport could almost equally well be used on land. The Air Cushion Vehicle (ACV), better known as the hovercraft, first came into operation in 1959. It weighed four tons, carried three men and skimmed across the sea at 25 knots.

An English inventor, Christopher Cockerell, had noticed how a strong downward draught of air could raise an object from the ground. He reasoned that in this position the object could be moved forward by additional propellers.

Cockerell carried out his first experiments with an air blower and a coffee-tin. He pushed the nozzle of the vacuum cleaner pipe through a hole in the tin, reversed the airflow so that it blew instead of sucked, and the tin was raised off the ground.

A hovercraft is equipped with extremely powerful motors which blow a very strong down-draught. This draught raises the craft above the surface of the water and large propellers move it along. Hovercraft have been used all over the world since the 1960s. They can travel at up to 70 mph (110 kph), carrying scores of passengers and vehicles.

Lighter than Air

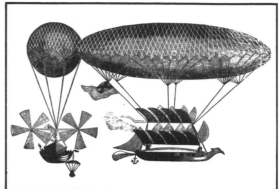

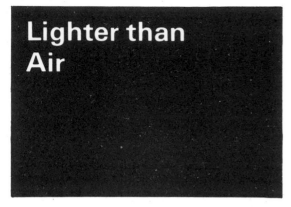

Man's first ascent into the upper atmosphere literally went up in smoke. It took place on November 21, 1783, in Paris in a hot-air balloon which floated across the French capital.

Before that flight, however, there had been many ideas, some of them weird and wonderful, as to how human beings could travel through the air.

During the Middle Ages, the philosopher Albertus Magnus wrote about a device which basically resembled a balloon.

Roger Bacon, the English monk and pioneer scientist, thought that a globe made of thin copper and filled with air might fly.

In 1670 a Jesuit priest called Francesco de Lana designed a craft that resembled a boat having a single mast fitted with a sail. The mast stood on the centre of the boat and was surrounded by four hollow spheres placed on top of poles.

De Lana thought that if the air was extracted from the spheres, the ship would float upwards and be blown along by the action of the wind on the sail. What De Lana did not realize was that the spheres would have been crushed by the pressure of the air on them.

During the 1700s, two Frenchmen, the brothers Jacques and Joseph Montgolfier, noticed that smoke from a fire always rose and floated in the air. From this they reasoned that if the smoke could be captured, it could be used as a lifting force.

They first experimented by filling paper bags with smoke. They burned wood and straw to obtain a really smoky fire, believing that the denser the smoke, the greater the lifting power.

On June 5, 1783, the Montgolfiers tested a huge cloth bag 35 feet (11m) in diameter which they filled with smoke. The primitive balloon actually rose.

Three months later they attached a basket to a smoke-filled bag, and a duck, a sheep and a cockerel became the first living creatures to make a lighter-than-air flight. The animals landed unharmed after an eight-minute voyage.

The first free aerial voyage by human beings took place when Pilâtre de Rozier and the Marquis d'Arlandes volunteered to go up in a Montgolfier balloon equipped with a brazier which burned wood and straw. The dauntless pair carried a bucket of water and sponges in case the fire got out of control. The two men ascended 500 feet (152m) and stayed airborne for 25 minutes.

Hot-air balloons are still used for sport today.

The Montgolfiers soon discovered that it was not the smoke which lifted them, but the hot air from the fire – indeed, modern balloonists use a flame-thrower to inject hot air into the gas envelope. From this discovery other experimenters reasoned that a lighter-than-air gas would do as well or better.

Shortly after the first Montgolfier flights, the French physicist Jacques-Alexandre-César Charles invented a

balloon made of rubberized silk inflated with hydrogen. Charles and a companion made the first hydrogen-balloon ascent on December 1, 1783. The balloon, which travelled some 25 miles (40 km) from Paris, was very much like a modern balloon. The gas envelope was held inside a net. The aviators carried stones as ballast. They could also partly control the balloon by operating a valve in the envelope.

Charles took along a thermometer and a barometer. He discovered that the temperature fell at altitude. From the barometer he was able to calculate the altitude from the falling air pressure.

Balloons similar to the one made by Charles were used during the Napoleonic Wars in the late 1700s and the early 1800s, and in the American Civil War during the early 1860s. Their main military function was to observe the movements of the enemy.

Sometimes they were used for attacking. The first air raid in history was carried out by the Austrians against Venice in 1849, when crews dropped bombs from a balloon.

During World War Two these non-rigid balloons were used as barrages to prevent enemy aircraft from flying low enough to attack towns and shipping with accuracy.

Balloons reached the stratosphere long before other

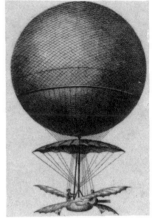

Above: The first ascent in a hydrogen-filled balloon, made by a French physicist, Jacques-Alexandre-César Charles, with Nicolas-Louis Robert, one of two brothers who built the balloon.
Above left: George Cayley, 'father of aeronautics', designed this balloon in 1837. It incorporated many of the features that became common in airships, such as the gondola and propellers.

Left: A model of Francesco de Lana's proposed airship. The weight of the copper spheres alone would have prevented it from flying.
Above: The balloon used by the French pioneer Jean-Pierre-François Blanchard for the first aerial crossing of the English Channel in 1785, accompanied by the US physician John Jeffries. The somewhat fanciful design was due to the imagination of the contemporary artist!

A balloon ascent of the 1880s, part of a fête held to celebrate an election victory.

The 'Graf Zeppelin', the most successful airship ever built. In the late 1920s she made two transatlantic crossings and a round-the-world trip.

forms of aircraft. On May 27, 1931, the Swiss balloonist Auguste Piccard ascended to 51,775 feet (15,780m) in a balloon fitted for the first time with a pressurized cabin. On May 4, 1961, M Ross and V Prather of the United States set up a record of 113,739 feet (34,667m) over the Gulf of Mexico.

Non-rigid balloons are difficult to control and steer. It was not until rigid and semi-rigid balloons were invented that lighter-than-air travel became more practical. These balloons became known as airships or dirigibles (able to be directed or steered). The gas envelope was supported by a metal framework.

The first successful dirigible was designed by the French engineer Henri Giffard in 1852. It was a cigar-shaped balloon from which hung a basket. The airship was moved by a steam engine which generated three horsepower and travelled at 5 mph (8 kph). A rudder was used to steer it. The airship flew 17 miles (27 km) on its first flight, but the engine was not powerful enough to allow the craft to be turned in a breeze.

La France, an airship designed by Charles Renard and A C Krebs in 1884, was powered by an electric motor and reached a speed of 14 mph (22kph).

David Schwarz, an Austrian engineer, built the world's first rigid airship. It crashed during its maiden flight in 1897.

Schwarz's work inspired Graf (Count) Ferdinand von Zeppelin, a retired German army officer. Zeppelin invented an airship in which light metal girders supported the gas envelope. Zeppelin launched his dirigible on July 2, 1900, at Lake Constance in Germany. The airship was driven by two 15-horsepower engines which worked two four-bladed propellers.

Zeppelin's airships operated the world's first airline from 1910 to 1914. More than 30,000 passengers were carried without a fatality. During World War One, Zeppelin's dirigibles could fly at 50 mph (80 kph), and were used to bomb Britain.

By the 1930s dirigibles measured more than 800 feet (240m) in length and 130 feet (40m) in diameter. The passengers and crew travelled in a cabin, called a gondola, underneath the hull. Power came from large diesel engines.

A famous airship of this type was the *Hindenburg*, launched in 1936. Its gas envelope held more than seven million cubic feet (198,000 cu. m) of hydrogen gas and the machine weighed 130 tons. It cruised at 78 mph (125 kph) and made 54 flights in all, 36 of them across the Atlantic. It caught fire and was destroyed during a storm at Lakehurst, New Jersey in the United States, killing 36 people.

The destruction of the *Hindenburg* followed similar disasters. In 1930 the British-made R 101 and in 1933 the American *Akron* had crashed, and it was obvious that airships were vulnerable to lightning and gales. Consequently, dirigibles were abandoned as a form of passenger transport.

During the 1970s, however, the use of airships as freighters for carrying cargoes was being re-examined. One possible use would be to carry space rockets from their storage sheds to the launching pads. The safer gas helium is generally used instead of the highly-volatile hydrogen.

The disaster that ended the major development of airships: the destruction of the 'Hindenburg' at Lakehurst, New Jersey, in the United States, on May 6, 1937, during a storm.

Heavier than Air

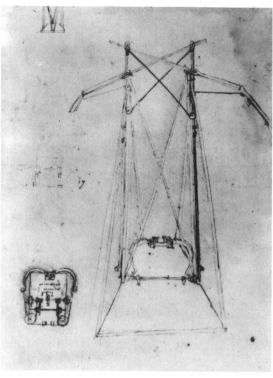

This was Leonardo da Vinci's original drawing for an aircraft with flapping wings, made back in 1485. The handwriting reads from right to left because Leonardo, who was left-handed, wrote that way.

The first true aeroplane was a strange-looking contraption of wood and wires which lurched into the air on December 17, 1903, at Kitty Hawk, North Carolina, in the United States. The pilot was lying in the middle of the machine, hanging on for dear life.

By the toss of a coin he had won the right to fly the machine and make the first sustained and controlled flight under power. His name was Wilbur Wright, and his co-inventor, and the man who lost the toss, was his brother Orville.

Flyer 1, as the Wrights called their machine, was the great-grandfather of the supersonic jets which speed to all parts of the world today. By their invention, the Wrights not only made Man's centuries-old dream of flying into a reality: they also effectively reduced the size of the globe.

There had been many previous attempts at heavier-than-air flight before the Wrights' success. As far back as 1325, children in Europe were playing with a toy fitted with windmill sails, which they set spinning by pulling at a piece of string. As the sails rotated, the toy rose into the air . . . the first helicopter.

In 1485 that Renaissance genius of many talents, the Italian Leonardo da Vinci, sketched a design for a machine which he called an ornithopter, meaning bird-winged. The machine's wings were to be worked by a man flapping them up and down as he lay along the wooden fuselage.

Leonardo's machine would never have flown. It was based on a misunderstanding of how birds fly. For centuries, men believed that birds fly by flapping their wings downward and then backward, rather like a swimmer doing the butterfly stroke.

It was not until the 1800s that scientists discovered that birds use only the inner part of their wings to stay airborne and to glide. The thrust which drives them through the air comes from primary feathers on the tips of their wings. These feathers twist backwards as the wing flaps down, and act as propellers. Modern aircraft use this principle. The wings keep the aeroplane airborne and the propellers or the jet engines provide the thrust.

Sir George Cayley, an English landowner, born in 1773, was one of the first men to discover this principle. A remarkable scientist, he invented the caterpillar track, the first lightweight cycle wheel, and an artificial hand with moving joints. Cayley is known as the Father of Aero dynamics because he wrote a book about the behaviour of objects in motion through the air.

In 1849 he built a triplane, a primitive aircraft with three wings, one on top of the other. He used a young boy as a test pilot. The triplane was made airborne by a squad of men pulling it down a hill like a kite. The triplane actually took off and flew for several yards before landing on its three-wheel undercarriage.

In 1853 Cayley sent his coachman on a trial flight over the dales near Scarborough in Yorkshire. Once more the aircraft flew for a short distance and then it crashed. The coachman was unhurt, but he gave in his notice, declaring he had been employed to drive horses and not to fly.

Cayley's machines were gliders, but he already knew that it was impossible to make a flying machine on the flapping wing principle. Cayley knew, too, that an aircraft would need certain controls to steer it and keep it stable. He also realized that it would need a propeller. The problem was: how was a propeller to be powered?

Many experiments followed. The English engineers William Henson and John Stringfellow made flying machines powered by steam-engines from the 1850s onward. These machines simply could not get off the ground because the engines were too heavy.

One of the earliest aviators was Otto Lilienthal, a German mechanical engineer. He had studied the flight of birds closely and written a book on the subject in 1889.

Lilienthal built and flew gliders with one and two wings – monoplanes and biplanes. They were like the hang-gliders of today. Lilienthal hung from them, gripping the structure with his hands and arms. To

And this is a reconstruction of Leonardo's flapping-wing flying machine, made for an Italian documentary film about the inventor's life.

Two historic gliders, the precursors of powered flight. *Above,* one of the last flights made by Otto Lilienthal. His glider bears a strong resemblance to today's hang-gliders. *Right,* one of the Wright Brothers' experimental gliders being launched.

Two historic flights: *Right,* the twelve seconds which revolutionized transport, with Orville Wright at the controls of the first powered aeroplane to fly, at Kitty Hawk, North Carolina, on December 17, 1903; *below,* Louis Blériot making the first cross-Channel flight from Calais to Dover on July 25, 1909, which won him a prize of £1000.

steer them, he swung his legs backwards or forwards or tilted his body.

By jumping from the top of a hill or a cliff, Lilienthal could glide up to 750 feet (229m). His flights inspired other pioneers of the air. He was killed one beautiful summer day when a sudden gust of wind unbalanced his glider and sent it crashing to the ground.

The Wright brothers were among those who studied Lilienthal's early flights. By the time they took an interest in aviation, the German engineer N A Otto had perfected the internal combustion engine in 1876. The Wrights, therefore, had a potential source of power for manned flight.

Internal combustion engines are quite heavy, but they generate far more power in relation to their weight than steam-engines. It was this extra power which made flight possible.

At first the Wrights confined themselves to carrying out their tests on gliders or with kites. Their first task was to devise some way of controlling a glider if it should be suddenly caught in the type of wind which had killed Lilienthal.

The brothers invented a method of warping, or twisting, the wings of their gliders. This warping made the wing present changed shapes to the air currents, and it thus became possible to gain greater control over their craft.

The Wrights' warp-wing invention was the forerunner of today's airflaps and ailerons. The Wrights also installed rudders in their gliders.

They wanted to use a motor-car engine, but it was too heavy. So they designed a lighter, water-cooled engine which developed 12 horsepower.

The brothers had also studied the problem of propellers, and had discovered that a propeller must have the same shape as an aircraft wing.

If you take an aircraft wing and slice across it, the section obtained has a curved upper surface and a flattened lower surface. As the wing moves through the air, its shape causes the air to press against the lower part, and move away from the upper part. The lift which keeps the aircraft airborne is thus provided by a combination of pressure from below and an even greater suction from above.

A section of a propeller is similar to that of a wing. The curved part is at the front and the flattened part is at the rear of the blade. As the blades rotate, the same lifting power is produced, except that it works in a forward direction, producing pull instead of lift.

The Wrights' first flight on that December day in 1903 lasted 12 seconds. The distance covered was 120 feet (36m), just about half the length of one of today's jet airliners. The Wrights made four flights that day. The last one covered 852 feet (260m) of ground distance. All the flights were sustained and controlled.

Legendary inventor

The first attempts to invent a flying machine undoubtedly took place in very ancient times, as witness the many legends that have come down to us. The most famous of these is the Greek tale of the inventor Daedalus, who offended King Minos of Crete and made himself wings of wax to fly to safety. He made another pair of wings for his son Icarus, who flew too near the Sun and crashed when his wings melted.

Daedalus is also credited in legend with inventing the axe, the saw, the gimlet, the plumbline – and glue!

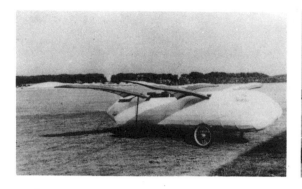

Two experimental aircraft of the years immediately before the First World War: *Far left,* the Passat Ornithopter, one of several attempts to make a flapping-wing plane; *left,* the Lee-Richards Annular plane, a biplane with two circular wings. The examples shown here are reconstructions.

The Wrights also invented the first practical aeroplane. They called it Flyer III. It could bank and turn, perform a figure-of-eight manoeuvre, fly in a circle and stay in the air for half an hour at a time.

The next development in aviation was the monoplane, and the first successful aircraft of this kind was designed by the Frenchman Louis Blériot. On July 25, 1909, Blériot flew across the Straits of Dover from France to England. His aircraft was fitted with an engine of 25 horse-power. It travelled at 45 mph (72 kph) at an altitude of about 250 feet (76m).

Flying machines underwent enormous development during World War One. Biplanes, triplanes and monoplanes fought the first aerial battles. The first bombers were produced, including the Italian Caproni Ca 46 which carried a crew of three and had three engines. After the war, the Caproni continued service as an airliner carrying eight passengers.

In 1923 a Spaniard, Juan de la Cierva, invented the autogiro, an aeroplane without wings. It had an engine at its nose and a rotor above the fuselage on a vertical shaft. The nose-engine provided the forward thrust and the rotary wing was set in motion as the aircraft travelled along. The spinning rotor provided the necessary lift to keep the autogiro airborne.

The autogiro had to make a run to take off. It could land almost vertically, but it could not hover like a helicopter, which depends for lift and thrust on its rotary wing.

A German engineer, Focke-Achgelis, invented the first practical helicopter in 1936. It had twin rotors and it could rise and land vertically. It could also hover and tilt from side to side.

Unlike the autogiro, the helicopter's rotors are powered and the blades have a variable pitch so that the angle at which the blades 'bite' into the air can be varied by the pilot. By changing the pitch of the blades, the pilot can perform all the manoeuvres necessary. A smaller rotor at the rear of the helicopter helps to steer the machine, which would otherwise spin round in the air.

By the late 1930s aircraft were flying at speeds of up to 400 mph (640 kph). Airlines in the United States

The story of the helicopter began, like that of so many other inventions, with the Renaissance genuis Leonardo da Vinci. He left drawings from which this model was built.

A reconstruction of the first autogiro, invented by the Spanish engineer Juan de la Cierva in 1923.

The principle, proved by Hero and demonstrated in rockets, remained a novelty until the 1920s, when engineers began to experiment with jet propulsion as an alternative to the ordinary internal combustion engine. Frank Whittle, a British aviation engineer, patented a jet engine in 1930, and a Frenchman, Réné Leduc, produced a jet-powered model aircraft in 1938. But the first jet-powered plane to fly was German, the Heinkel He 178, which took to the air in 1939. In 1941, the British produced the first practical turbo-jet aircraft. A turbine was used to compress the air in which the fuel was burned.

Jet engines were modified in military aircraft to shorten take-off distance. The British-made Harrier 'jump jet', designed as a vertical landing and take-off (VTOL) plane, was first tested during the 1960s. The thrust from the engines can be swivelled downwards for vertical take-off or landing and turned to a horizontal position for forward flight.

As air travel expanded and the skies became more crowded, further inventions emerged to aid navigation. Besides radar (see page 24), these devices included the non-directional beacon (NDB). This beacon sends out a radio beam which is picked up by the automatic direction-finder (ADF) or radio compass aboard the aircraft. From this beam the pilot can plot his position.

The very high frequency omnidirectional radio (VOR) is a development from the NDB. The VOR sends at 360 beams like the spokes of a wheel, corresponding to all the angles of a circle. These beams are also picked up by the aircraft, but over a wider band than on the NDB.

A further development is the Distance Measuring Equipment (DME), which works on radio pulses sent out by an aircraft. The pulses trigger off a response from the DME on the ground, and the pilot knows how far away he is from that point.

Above: The first jet aircraft to go into service in World War Two, the German Messerschmitt Me 262 fighter.
Right: The Rolls-Royce 'Flying Bedstead', prototype of all vertical-take-off aircraft, undergoing trials.

Far right: A Westland Sea King helicopter lowering a sonar buoy into the water during an anti-submarine detection exercise.

and Europe were carrying passengers, and air travel was becoming more widespread.

The invention of the jet engine greatly increased the speed and the altitude at which aeroplanes could fly.

Hero, a Greek scientist, designed the first jet-driven mechanism about 2000 years ago. He used the rapid ejection of steam to make a ball rotate.

A jet engine works by burning fuel rapidly so that it produces hot, expanding gases. The gases push against the interior of the combustion chamber, and rush out at the opening at the back. But it is not the action of the jet on the outside air that moves the jet forward, but the interior pressure against the rest of the engine. For this reason, a jet engine works best in the vacuum of outer space, because there is no air outside to counteract the pressure of the escaping gases.

A modern swing-wing military jet fighter, the MRCA prototype 02. Changing the plan of the aircraft can give greater control at low speeds for landing and take-off.

Space

The dawn of the Space Age was ushered in by an odd little sound – 'bleep bleep'. This was the radio signal emitted by *Sputnik 1*, a Soviet invention and the first artificial satellite to be launched into orbit round the Earth. *Sputnik 1* was launched on October 4, 1957, and it created a sensation.

The rocket, the main vehicle for space travel, was probably invented by the Chinese. There are historical records of 'fiery arrows' being used in a battle in AD 1232. Some 50 years later, rockets were being used as weapons of war in Europe.

A Russian schoolteacher, Konstantin Tsiolkovsky, correctly explained the principle of rocket flight in 1903, but the first scientific experiments with rockets were carried out by the American Robert H Goddard. Goddard built the first liquid-fuel rocket in 1926, and in 1930 he launched a rocket that zoomed to 2000 feet at 500 mph (600m at 800 kph).

Most high-altitude rockets use liquid fuel. Solid fuel consists of mixtures of resin, rubber or asphalt with an oxygen compound. Liquid-fuel rockets use liquid oxygen to produce combustion of such materials as alcohol, peroxide, aniline or nitric acid.

Further experiments with rockets were carried out during the 1920s and the 1930s by a group of German scientists led by Hermann Oberth.

By World War Two, rockets had become so developed that they were used as major weapons in land, sea and air battles. Rocket guns were fitted to aeroplanes and the Soviet army used rocket artillery to devastating effect against the German invaders.

Germany launched liquid-fuelled rockets against Belgium and Britain. These V-2 rockets carried powerful warheads packed with explosives, and were guided by radio to their targets.

After the war, German rocket experts, headed by Wernher von Braun, went to the United States and continued their experiments. The first programmes to explore the Earth's upper atmosphere and the threshold of outer space took place from 1946 to 1951. The United States launched some 60 V-2 rockets from New Mexico. The rockets carried instruments which brought back valuable information about conditions high above the Earth.

Soviet rocket experts were also carrying out similar experiments. In 1957 a Soviet rocket carrying a heavy payload reached an altitude of 294 miles (473 km).

By the end of the 1950s, American and Soviet scientists had evolved many intricate instruments and machines for space exploration, to measure radiation and note the behaviour of rockets travelling at high speeds. The computer was also developed to calculate speeds and trajectories necessary for placing an object into Earth-orbit.

Sputnik 1 marked the success of this programme. It was soon followed by another major step forward. On November 3, 1957, a small black and white dog called

Laika became the first living creature to travel in space. The Americans launched their first artificial satellite, *Explorer 1*, on January 31, 1958, and the race was on.

The Soviets produced a space capsule large enough to carry a human being, and Yuri Gagarin became the first man to go into orbit round the Earth on April 12, 1961.

As the number of manned spaceflights grew, the transporting vehicles became more elaborate. The Americans were the first to develop craft which could link together in space. The first space-link took place on March 16, 1966, opening the way for the assembly of large units in space.

One of the most important developments of this programme was the American *Skylab*, an assembly, 118 feet (36m) long, of spacecraft fitted with telescopes, cameras and other specialized equipment. It provided a unique opportunity to study the stars and the planets.

On February 8, 1974, the *Skylab* astronauts set up a record stay of 84 days in space.

Top: Since US astronaut Edward White made the first space-walk from Gemini 4 in June 1965, such extravehicular activity, as it is called, has become a commonplace of space flight.
Above: The Russian space probe Venera 7, which soft-landed on Venus in 1970 and sent back the first information from the surface of the planet.

The Russian Luna 16 automatic probe, which was invented to dig lunar rock samples from the Moon's surface and fly them back to Earth in 1974.

Far right: The Russian spacecraft Vostok I in which Yuri Gagarin made the first manned space flight on April 12, 1961.

Right: The Russian self-propelled vehicle Lunokhod 2, which explored the surface of the Moon in 1973 while controlled from Earth.
1, directional antenna;
2 and 8, TV cameras;
3, photoreceptor;
4, solar battery;
5, magnetometer;
6, angular reflector;
7, astrophotometer;
9, instrumentation pack;
10, telephotometers;
11, vehicle's roadability gauge.

The American Apollo 15 Moon landing included a battery-powered Lunar Roving Vehicle with which the astronauts explored a wide area of the Moon's surface. In this picture James B Irwin is saluting the US flag, with the lunar module 'Falcon' in the background and the LRV on the right.

Meanwhile, there had been important developments in the field of satellites. In 1962 the United States launched *Telstar*, the first of a series of communications satellites. Placed in orbit at a speed equal to the rotation of the Earth, *Telstar* hovered over the middle of the Atlantic Ocean and acted as an extremely lofty aerial for radio and television signals. Live television transmissions across the two hemispheres became possible for the first time.

In 1963 the United States launched *Tiros 7*, a weather satellite. *Tiros* continuously circled the Earth sending back pictures of cloud conditions and air currents. For the first time, meteorologists were able to observe huge hurricanes building up and were able to give early warning of these.

Tiros was the first of the 'spy' satellites, which can detect and photograph the movements of troops and the build-up of armaments and military depots. The photographic equipment on them has became so highly developed that they can photograph a watch on the wrist of a man 200 miles (320 km) below.

Surveyor-type satellites were invented which can detect mineral resources in the Earth through the use of light-spectrum photography. Scientists can tell from the colours produced on the photographs what type of vegetation and minerals may be found on particular parts of the planet.

The major target remained the Moon. In 1966, Soviet engineers built the first spacecraft to make a soft landing on the Moon. For three days, the machine sent back to Earth pictures of the lunar surface.

Another Soviet invention was the *Lunokhod*, or moon-walker, an eight-wheeled vehicle looking like a motorized bathtub. *Lunokhod* travelled more than 6 miles (10 km) across the dusty wastes of the Moon, beaming television pictures back to Earth.

The Americans soon produced an equally grotesque but useful invention, *Surveyor 3*, a spidery vehicle which made a soft landing on the Moon in April 1967. It spent 18 hours scooping up samples of lunar soil, analyzing them and radioing its findings to Earth.

Meanwhile, the Russians were looking farther into space. In October 1967 they sent the rocket *Venera*, to drop a capsule of instruments on Venus. The instruments sent back information indicating that the planet's atmosphere was chiefly made up of carbon dioxide.

In preparation for their manned expedition to the Moon, the Americans invented the Lunar Module, promptly nicknamed the 'ugly duckling'. An ungainly, spindly craft, it was fitted with 18 engines and weighed 16 tons. On May 22, 1969, it descended in a controlled rocket flight to less than nine miles (14 km) above the Moon, carrying two men from the Apollo spacecraft.

On July 20, 1969, a similar ugly duckling transported Neil Armstrong and Edwin Aldrin to the first human landing on the Moon.

In the 1972 Moon expedition, American astronauts demonstrated the latest invention in space travel: the Moon-car. It had broad wheels and special suspension to enable it to travel along the dusty surface of the Moon without getting stuck.

One of the astronauts accidentally knocked off a mud-guard and the travellers were covered in masses of moondust. The car, powered by electric batteries, was left 'parked' on the Moon.

Another development in space transport concerned the problem of Man's re-entry into the Earth's atmosphere after a spaceflight. The Earth is surrounded by a dense blanket of air which sets up friction resistance when a space capsule penetrates it at 25,000 mph (40,000 kph). This friction creates temperatures of up to 5000°F (2750°C). To meet this problem, scientists produced new plastic surface coverings for the returning capsules. These plastic materials were designed to burn and take the heat in a process known as ablation. The temperature inside the capsule remained at 70°-75°F (21°-24°C).

HISTORY OF PARACHUTES

The parachute was probably invented by Leonardo da Vinci about 1500, though there is no record of his having tried it out. Like so many of Leonardo's bright ideas, this one went into oblivion, to be re-invented in 1783 by a French chemist, Louis Lenormand. In that year he leaped bravely from a tower at Lyon, France, with an umbrella five feet (1.5m) across in each hand – and lived to tell the tale.

Two years later a pioneer French balloonist, Jean François Blanchard, dropped overboard a dog in a basket to which he had attached a small parachute.

Blanchard later claimed to have taken the same plunge himself in 1793, but credit for the first man to parachute from a balloon is generally given to another French aeronaut, André Garnerin, who made his first descent in 1797. He used an umbrella-shaped parachute, about 23 feet (7m) across, made of stout cloth.

Parachutes changed very little in design until the 1930s, apart from the provision of a small vent in the top of the canopy. Such parachutes had the disadvantages that they opened with a jerk, and were liable to swing the unfortunate parachutist to and fro. In 1938, German designers came up with the ribbon parachute, in which the panels of the canopy are not completely joined; this design allows a slight flow of air through the canopy and makes for a steadier descent.

Modern parachutes are of many designs, some of them remarkably similar to the pioneer glider of Otto Lilienthal, illustrated on page 57. Special guide panels, controlled through lines by the parachutist, enable the user to control the direction of his fall with considerable accuracy. In sport parachuting, which achieved international status in 1951, competitors are expected to land within a circle only 7m across.

A French cartoon recalling the adventure of a showgirl named Mrs Graham, who fell out of a balloon 30m above the ground in the 1830s and was saved when her skirt acted as a parachute.

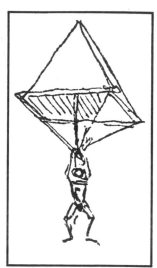

Above: The principles of the parachute were first appreciated by Leonardo da Vinci in the 1400s.

Above: Another pioneer of parachute design — though it seems unlikely he ever tried it out — was the Hungarian mathematician Fausto Veranzio, who produced this design while he was living in Venice.

The French aeronaut André Garnerin made this spectacular parachute descent in 1802, an exhibition jump on 2400m from a balloon.

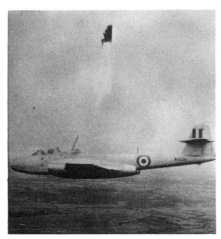

Above: An ejection seat of the kind commonly fitted to combat aircraft. It hurls the user clear of the plane and its slipstream.
Above centre: An airman a split second after ejecting from an aircraft. A separate parachute brings the seat down to earth.
Above right: An automatic parachute release can be set to operate at any height desired. It will open a free-fall parachute in the event of the wearer failing to pull the rip-cord.

Above: An army vehicle on three heavy-drop parachutes descends towards a dropping zone.
Above centre: A mass drop of 900 men during a NATO exercise.
Above right: One of the Royal Air Force's 'Red Devils' team during a demonstration jump. Parachutes of this kind are designed for maximum ease of manoeuvre.

The parachute as an aid to braking: an RAF 'Lightning' is pulled to a halt.

Agriculture

Man began farming about 10,000 years ago, in the region of Turkey and Palestine. Later, agriculture spread to Egypt and Iraq, and it appears to have developed independently in China and the Americas. Since then, it has spread to most parts of the Earth. At first, farming was largely a matter of human muscle-power, but gradually inventions led to the use of animals and eventually other sources of power, so that the time taken to produce a given amount of food has been reduced dramatically.

Early farmers used sticks and hoes to stir up the ground and break up lumps of soil. Such methods are still used in some primitive places. Ploughs were first used by the Sumerians of Mesopotania and by the Egyptians about 3000 BC. The Sumerian plough had a triangular share, or blade, of wood or bronze, mounted on a frame which a man pushed over the ground. Then the Sumerians tamed oxen to pull the plough, and designed a yoke to harness the animals' power. The oxen could pull heavier ploughs, which cut deeper into the soil.

Horses were first used in agriculture around 2300 BC. Their efficiency was greatly improved by the invention of the horse-collar. The Babylonians, who followed the Sumerians about 2000 BC, devised a seed drill. The drill was a hollow wooden tube, packed with seed, and pulled behind the plough. The seed trickled out as the ground was prepared. The Chinese seem to have made a similar device around 2800 BC. Without a drill, the seed was broadcast on the ground and lightly covered by hoeing.

Important additions to the plough were the coulter, a vertical blade mounted in front of the share – in use in Roman times – and the mould board, a curved piece of wood that turned the soil over when it had been cut. A mould-board plough was patented in England in 1730.

Far right: Farming in ancient Egypt – the two digging tools in the foreground are models based on tools of about 2000 BC, while the ox-plough is a copy of one illustrated on a monument of about 1500 BC.

Right: Ploughs developed only slowly between the days of ancient Egypt and medieval Europe. This carving on the campanile of the cathedral in Florence dates from the 1300s.

Opposite: Hay-making with the 'Iron Lady', a cage-like machine used for making small hay-ricks.

Although iron was in use from 1500 BC onward, it served mainly to provide a hard-wearing edge to the share. In Europe, ploughs were made of wood well into the 1700s, and the early American pioneers hacked their new farms out of the virgin forest with wooden ploughs.

Robert Ransome, an English agricultural tool-maker, produced the first all-cast-iron plough in 1785, and made a self-sharpening version in 1803. Even this plough gave trouble on heavy clay soils, which stuck to cast-iron. But in 1830 John Deere, an Illinois blacksmith, invented the steel plough, which was much more efficient.

In the early 1700s, a variety of implements were developed to prepare the soil and cover the seed. The most important inventor of this period was Jethro Tull, an English farmer, who invented a type of seed drill in the 1730s which sowed the seed in rows below the surface. This device saved a lot of seed. Tull encouraged the use of cultivators and harrows to break up the soil after ploughing. There were disc cultivators, with a number of parallel metal discs to break up the clods, and tyne cultivators, with curved metal bars, pointed at their lower ends.

The harrow was merely the branch of a tree when it was first devised as a method of breaking up soil. But by 1850 metal harrows were in use, and an all-steel one was developed in Norway. Use of these implements meant that only one ploughing was needed.

As soon as the steam engine had been developed for transport, it was applied to farming. Steam ploughing was first done in the 1860s. The early traction engines which were used were heavy, and tended to compact most soils, so their use was limited. But two stationary steam engines on opposite sides of a flat field could pull implements to and fro on a chain running between them.

The horse and ox as prime sources of agricultural power were finally challenged by the internal combustion engine. The first farm tractors using engines of the kind were produced in 1892 by an Iowa blacksmith, John Froelich. During the next 50 years, tractors almost completely replaced horses and oxen in Europe, North America, Australia and many other parts of the world. In 1918, the International Harvester Company of the USA made tractors which could provide power for operating other machines, as well as for hauling ploughs, and they meant the end of steam engines, even for standing operations.

Plants need water and food to grow. The ancient

This picture of ploughing from an 11th-century calendar shows four oxen dragging the cumbersome tool through the ground. In heavy clay, as many as eight oxen would be needed.

Far left: Many varieties of cultivators have been invented for breaking down the ploughed soil. This one has hook-shaped tines. *Left:* The Spray Gapper is one of many inventions designed to make the farmer's work easier.

Egyptians were lucky; every year the Nile flooded their lands, and also put fresh fertile soil on them. But not enough land was fertilized in this way, so more land was brought into use by irrigation, using sluices and ditches. Irrigation frequently involves lifting water to higher levels, and devices for serving this purpose have kept inventors busy for centuries.

One of the simplest such devices is that known in Egypt as the *shadoof,* and in India as the *denkli.* It consists of a pole pivoting on an upright post and mounted closer to one end than the other. A heavy counterweight is fixed on the short end. The operator hitches an empty bucket on the long end and pulls the pole down; the counterweight pulls the full bucket up. This device was used in ancient Egypt, and is still to be seen on the banks of the Nile.

Another ancient device is the Archimedean screw, said to have been invented about 200 BC by the Greek

The Ivel tractor of 1902 was the world's first successful medium-powered internal-combustion-engined farm tractor. It was designed by the British engineer Dan Albone after five years' work.

Jethro Tull invented his seed drill in the 1730s.

The Archimedean screw is still used in some parts of Egypt to raise water for irrigation.

mathematician Archimedes. It consists of a screw turning inside a cylinder – somewhat like the business part of a domestic mincing machine. The cylinder is placed at an angle, so that one end dips into the water. The operator turns a handle, and 'screws' the water up. This device, too, is still used.

Equally ancient is the Persian Wheel, known as the

Barbed wire

Some inventions have results far beyond their creator's dreams. One of these was barbed wire, which changed the whole character of the Wild West in a few years.

The American West was won by tough pioneers, some of them prospectors, some farmers who tilled the soil, and some of them cattlemen. The Great Plains country of the Far West became cattle country in the years immediately following the American Civil War, which ended in 1865. Cowboys drove their cattle to graze on the wide grasslands, much of which was unclaimed territory. When the beasts were fattened, they were rounded up and driven to the nearest railheads and markets.

But US government policy was to encourage permanent settlers, men who would till the soil and make it fertile. Clashes developed between the cattlemen and the farmers when the prairie was fenced in, and the free grazing lands gradually shrank. In any case, fencing was difficult and expensive. Wood had to be hauled in at considerable cost, plain wire fences were broken down by the cattle, there were not enough rocks to build stone walls, and hedges just took too long to grow.

The most successful of the hedges were grown from prickly shrubs such as osage orange, which has spikes that cattle dislike. It was probably the memory of the cattle's reaction to such prickly bushes that prompted an unknown salesman to experiment with a prickly fence: along a length of wire he hung a strip of wood with spikes sticking out of it. And he demonstrated it one day at the county fair at De Kalb, Illinois, on the edge of the prairie.

Local farmers looked on with interest as the salesman demonstrated how cattle would shy away from the spikes. Among them was Joseph Glidden, an elderly man who went home in a thoughtful frame of mind. Surely, he thought, there must be a simpler and cheaper way to achieve the same effect. After some experiments, Glidden hit on the notion of making a fencing wire of two strands of ordinary wire twisted together, with short wire spikes every so often along the twists. Barbed wire was born.

That was in 1873, and at first Glidden made his barbed wire by hand in his own backyard, using his grindstone to twist the strands of wire together. The new fencing material – light, tough, cattleproof and windproof – became popular almost overnight. Within ten years it was produced by machinery at the rate of 60 miles (97km) of barbed wire every hour.

The wire led to all sorts of trouble. Two other De Kalb farmers, Jacob Haish and Isaac Ellwood, had similar ideas at about the same time. Although Glidden had the first patent, there were years of lawsuits before his claim was finally substantiated. And out on the range there was often open warfare between the cattlemen, anxious to keep the range free, and the farmers who fenced in the land they had bought from the government and were anxious to farm. There was even warfare between rival cattlemen, who used barbed wire to enclose waterholes or barricade cattle trails. In 1899, barbed wire began to play its part in real warfare, in the bitter Boer War in South Africa.

sakia in Egypt. A vertical wheel turned by oxen raises an endless chain of filled buckets which automatically tip into an irrigation channel.

Sometimes the land holds too much water, and has to be drained. An important invention of the late 1700s was the mole plough. This has a pointed metal cylinder – the 'mole' – mounted under a bar which is attached to a frame. The mole is drawn through the soil to make a channel for the water.

From earliest times, human and animal manure was spread on the soil. During the 1700s it was found that saltpetre (potassium nitrate) greatly improved plant yields. In the early 1800s, guano (bird dung) from tropical islands, basic slag from the steel industry, and ground-up bones were also applied.

Chemists discovered that plants could live on a diet of nitrogen, phosphorus and potassium, with smaller amounts of calcium and other elements. Soluble salts of these elements were then mixed in the right proportions to make artificial fertilizers. To use these fertilizers. efficiently, American farmers invented the combine drill in the 1840s This drill sowed the seed and deposited its correct food supply close at hand.

Soil is not essential for growing plants, provided they have enough water, sunlight and balanced food. In the hydroponic system, invented around 1860, plants are grown in sand or gravel through which water percolates, or they float in a stream of water. Dissolved in the water is all the food the plants need.

The harvesting of crops has also taxed Man's inventiveness to the full. The first sickle was made by

Stone-Age Man more than 10,000 years ago; it consisted of a piece of deer antler, with sharp bits of flint set into it to provide a cutting edge. The origins of the scythe, a two handled tool, are almost as old. Scythes and sickles were used for cutting grain until the mid-1800s.

To separate grain from its outer husk, called chaff, and from the straw stems, farmers used to lay it out on a hard earthen floor and beat it with flails – short sticks hinged to long handles – or trample it by the hooves of oxen. The straw was then removed by hand, and the grain and chaff was tossed into the wind, which blew the light chaff away while the heavier grain fell back to the ground.

The first combine-harvester was made in Roman Gaul. It consisted of a cart pushed through the standing cereal by a horse or an ox, with knives on the front to cut off the ears of corn and a device to knock the grain out of the ears. The grain fell into the cart and the chaff blew away in the wind. This implement presumably had its disadvantages, for it was not generally adopted, and it was not until the 1800s that efficient machines for cutting and separating the grain were developed.

Joseph Boyce, an English inventor, patented a

Above left: Where water or wind power was not available, machinery was horse-driven. This – literally – 4 hp threshing machine dates from the early 1800s.
Above: Patrick Bell, a Scottish minister, invented this reaper in 1826.
Left: The 'Sila-Masta' not only makes hay straight into silage, but loads the silage into a lorry, which it draws along behind it.

Above: Before the advent of tractors, steam traction engines were used to power farm machinery.
Left and below left: Windmills of the late 1500s in a contemporary woodcut, and two mills still standing in a Spanish landscape as they did in the days when Cervantes wrote about Don Quixote.

Above: Modern combine harvesters can even cope with a flattened crop when wind and rain have done their worst with the grain.

Right: Milking the modern way, with a machine, and below, the Marchfell machine of 1889, one of the pioneer devices.
Bottom: Churning butter by hand, as it was done in the 19th-century and earlier.

mills, because the four sails were fixed to a shaft connected by gears to a central post. In this kind of mill there was a brake wheel, operated by a lever, to stop the sails, and a tailpost to turn the sails into the wind. The main post is held by struts and cross-timbers, and is enclosed by a timber wall to protect the stored grain and flour.

The tower mill or smock mill followed during the 1300s. In this type of mill, the sail shaft is fixed to a rounded cap which runs on a circular track round the top of a tower. In this way, the support and weatherproofing are combined.

In 1745, an English millwright, Edmund Lee, invented the fantail. This is a miniature windmill mounted at right-angles to the sails, and the action of the wind on it keeps the sails turned into the wind.

The sails were originally of canvas on a wooden frame, but they had to be furled and unfurled like a ship's sails, a hazardous job. In 1722, Andrew Meikle invented shutters like venetian blinds, that could be adjusted without having to climb along the sail.

In the 1800s, steam power came into use, and the quern stones were replaced by rollers of grooved iron or steel. The first large flour mill using these methods was opened in Budapest in 1840.

Among many inventions for the collection of animal products, the most important is the milking machine. Milking machines rely on a vacuum pump to suck the milk out of the cow, cups to fit over the teats, and a pulsator to apply vacuum in a rhythmic fashion like the suckling of a calf. These methods were first applied to milking between 1851 and 1895, and the use of milking machines gradually replaced hand-milking during the first half of the 20th century. At first the milk was drawn into buckets, but now it is passed into a vessel to be weighed, and then through a cooler to a tank.

It is said that the automobile 'king' Henry Ford once visited the Chicago stockyards. He noticed how the cattle were processed in the slaughterhouses and the carcases were carried to the workers, each of whom dealt with one stage of the process. He applied these methods to the making of cars, and from there mass-production has spread throughout industry – an example of how one invention leads to another, even in a different field.

mechanical reaper in 1799, the start of a long line of such inventions. The first two really successful reaping machines were those made by a Scottish minister, Patrick Bell, in 1826, and a Virginian engineer, Cyrus McCormick, in 1831. An equally successful machine was patented in 1833 by Obed Hussey, an engineer from Maine. A series of inventions followed to produce by the 1880s the first harvester which not only reaped the corn but also bound it into bundles ready for stacking.

The first successful threshing machine was the work of a Scot, Andrew Meikle, in 1786. From this it was only a step to re-invent the combine harvester. American inventors were to the forefront here, and the first modern combine was built in Michigan in 1836. Combines were hauled by traction engines or tractors until 1908, when a self-propelled combine was made in the United States.

Grain can be fed directly to animals, but for humans it is usually ground into flour. Stone-Age Man ground corn in a flat or slightly dished stone against which another was rubbed. Later people used a cup-shaped mortar with a club called a pestle. Grinding was mechanized with the invention of watermills and windmills.

The watermill was developed in the Roman Empire. The wheel has a series of paddles around its rim. For the undershot wheel, these paddles dip into a stream of water which has been channelled to the width of the wheel. Undershot wheels have an efficiency of about 30 percent. In the overshot wheel the water is poured on near the top, and the force of gravity on it is used to turn the wheel. Such wheels have 80-90 percent efficiency. Bevelled gears enable the wheel shaft to drive a vertical shaft. This carries the quern stones that grind the grain.

Windmills were first mentioned as used in Persia in AD 644. Europeans got the idea from the Arabs in the later Middle Ages. The first windmills were called post

Turnip top!

One very important invention in the history of agriculture was neither a tool nor a machine. It was a method, the so-called Norfolk or Four-Course system of cultivation.

The system was invented probably during the 16th century, but it was not widely known until the activities of an English statesman, Charles, Viscount Townshend. Townshend had a busy political life, for 11 years of which he was responsible for British foreign policy. Eventually he resigned and settled down to improve his estates in eastern England, at Raynham, Norfolk. He adopted the Four-Course system and popularized it. The system consists of growing in each field wheat or oats the first year, oats or barley the second, clover, swedes, kale or similar crops for cattle feed the third year and turnips the fourth year.

Townshend's friends nicknamed him 'Turnip Townshend' because of his enthusiasm. But his methods are still practised today.

HISTORY OF HOVERCRAFT

It often happens that a man who is an expert in one field makes an important invention in a totally different one — and that the final result is remote from either. Which explains why, when an English radio expert turned to boatbuilding, people suffering from severe burns had good cause to be grateful.

The inventor was Christopher Cockerell, an electronics engineer. Tired of his work, he and his wife spent a legacy on a business building small boats. After a while, Cockerell began to think: Why won't boats go faster? The basic answer was friction. Cockerell, accustomed all his life to solving problems, decided to have a go at this one.

To reduce friction, you need a lubricant. Cockerell thought air might be the answer. After many experiments, he hit on the solution — a curtain of air right around the hull of the boat in the form of an annular, or ring-shaped, jet. To prove his theories right, Cockerell bolted a cat food tin inside an old coffee tin to make a simple form of annular jet, attached them to an air blower, and measured the amount of lift. It was enough to raise a boat clear of the water so that it would float over it. In Cockerell's mind, the hovercraft was born.

That was in 1955. A few months later, Cockerell was testing a working model, which behaved perfectly. But from then on Cockerell faced a series of frustrations. Shipbuilders said his device was an aircraft; aircraft manufacturers said it was a boat; everybody was interested, but no-one would do anything practical to help.

Finally a British government agency, the National Research Development Corporation, took up Cockerell's cause. The inventor needed cash to patent his hovercraft overseas: within hours of seeing Cockerell's plans the NRDC chief, Lord Halsbury, authorized the necessary finance. Halsbury and his advisers realized they were on to something really big.

The first full-sized hovercraft, the SRN1, was built and ready for testing by May 28, 1959, just 13 months after Cockerell walked into the NRDC office. The little craft, 30 feet (9m) long and 24 feet (7m) wide, was a complete success. Within a few days it crossed the Solent from mainland England to the Isle of Wight, landing on a convenient beach, and then skimmed right across the English Channel to France.

Cockerell had found himself crowded out of the development of his invention; when greater lift was needed to clear the waves found in the open sea, he suggested flexible skirts, only to have his idea pooh-poohed. Yet two years later, research by other scientists reached the same conclusion. By 1966, Cockerell's part in future developments of the hovercraft was merely a consultative one.

But the hovercraft has gone from strength to strength and a great many other developments of the 'hover' principle have since taken place. They include:

The hoverbed: A light fabric bed supported on a rigid frame. When the patient lies on the bed, warm, sterile air is pumped into long pockets which form a seal around the side of his body, while the fabric of the bed drops away beneath the rest of the body, leaving it floating on air. The hoverbed is used in the treatment of severe burn cases: it allows the injuries to dry up rapidly and heal much more quickly.

The hover-transporter: A heavy-load transporter working on the hovercraft principle is now used for moving bulky loads. The weight of, say, a 200-ton transformer carried on a wheeled vehicle is transmitted to a road or a bridge by the wheels. With, for example, an eight-wheeled vehicle, each wheel transmits 25 tons to a very small bit of road, a load sufficient to cause severe damage. The hover-transporter spreads the load and enables heavy weights to be carried over bridges which otherwise would not support them. By using made to measure skirts, two 50-foot (15m) oil tanks were manoeuvred on a cushion of air across to rail lines and a road to a new site in July, 1967. The principle has been used many times since then.

The hoverkiln: This device was used for a time in the manufacture of delicate china, which 'floated' through the kiln on a cushion of air.

The hovermower: Under various trade names, lawnmowers which ride over the grass on a cushion of air are proving of great value on banks and uneven ground, as well as stretches of level turf.

The device which began it all – the two-can assembly which Christopher Cockerell used to prove his theories back in 1955.

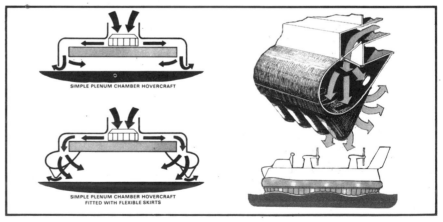

SIMPLE PLENUM CHAMBER HOVERCRAFT

SIMPLE PLENUM CHAMBER HOVERCRAFT FITTED WITH FLEXIBLE SKIRTS

How the hovercraft works: *Above, left:* The diagrams show how with flexible skirts more air is trapped beneath the hovercraft, giving greater lift; *above:* the skirt is a double-walled structure, the upper part a semi-rigid bag, the lower part a series of flexible extensions known as fingers, which stay in contact with the surface of the ground or water and form a complete air seal; *left:* how air is drawn in and distributed to give maximum lift.

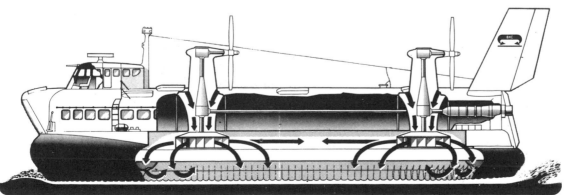

Right: Christopher Cockerell (in the cap) with the first model of the hovercraft.
Far right: The prototype model undergoing its first water trials.

Right: The first full-sized hovercraft, SRN 1, which was completed in May 1959.
Far right: The hoverbed, developed in 1967 and used for the treatment of severe burns.

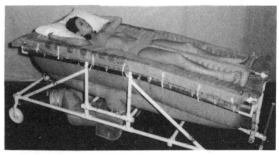

Right: The hovertrain, an experimental model tested in Britain but dropped. It was designed to run on a T-shaped concrete track.
Far right: The hover principle being applied to moving a heavy object — a 200-ton oil storage tank at Milwaukee, Wisconsin.

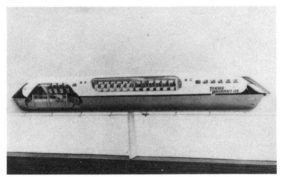

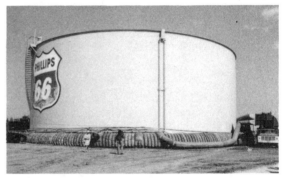

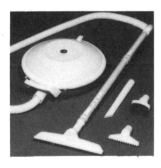

The air-cushion principle has been brought into the home, too. This vacuum cleaner moves effortlessly about the floor, riding on a cushion of air.

Two modern hovercraft ferries crossing the English Channel on regular service.

Opposite: The bow and arrow formed one of the earliest weapons used by Man. This tribesman from Nondugl, in the Central Highlands of New Guinea, uses his for hunting.
Far right: A Roman ballista in action, firing stone cannon-balls.

Many of today's weapons have developed from Man's first primitive efforts to pierce, stun or stab his enemies. The grenade is the descendant of thrown stones, cannon followed from siege guns, rifles from blowpipes, bayonets from spears and flame-throwers from Greek fire.

For thousands of years the infantry soldier's basic weapons remained much the same: bow and arrow, sword, dagger and spear. Changes evolved slowly and spread from one tribe or race to another as fighting men sought to deal one another more telling blows.

One invention towers above all others down the centuries, not only because of its effects on weapon development but also because of its influence on the course of human history. That discovery was gunpowder.

Nobody knows who invented gunpowder. The Chinese and the Indians have possibly the best claims to its earliest use, while the Greeks had a substance which blazed on contact with water and terrorized their enemies. 'Greek fire' was first used in AD 673. Saltpetre (potassium nitrate), the main ingredient of gunpowder, is first mentioned, however, by an Arabian scholar, Abd Allah, about 1200. Forty-two years later the English monk Roger Bacon published a formula for gunpowder and a German monk, Berthold Schwartz, developed its use in firearms in the 1300s.

Until then, the outstanding weapons were the various bows and siege engines and the breakthrough substance was metal. From 5000 BC copper, then bronze, then iron allowed fighting men to hone a finer edge on swords and achieve a better throwing and swinging balance on their weapons.

Until the Romans, the finest long-range weapons were bows and arrows. Nobody knows where they originated. They were used all over the world — by American Indians, African bushmen and pygmies, ancient Chaldean and Egyptian warriors, Asiatics and Europeans of the Middle Ages.

One of the most efficient of the early bows was the Asian composite bow, in which thin strips of wood sandwiched split horn, giving the bow a springiness which added many yards to its range. In tests held in the 1790s a composite bow was able to outdistance a wooden longbow, shooting more than 480 yards (440m).

During the Crusades the significant new weapon was the crossbow. Its advantage as a weapon was that,

once the bolt was set in its groove, it could be held there while the weapon was aimed and fired by trigger action. A later version, the arbalest, could pierce ordinary plate armour and had a range of up to 120 yards (110m).

The crossbow was always a somewhat clumsy weapon and much slower in action than the medieval marksman's finest tool: the longbow. Usually known as the English longbow, it probably originated in Wales, although similar bows were used by the ancient Egyptians.

The longbow was usually made of ash or yew, around six feet (1.8m) in length and with arrows about a yard (1m) long. It was capable of great accuracy and had tremendous penetrating power — it could pierce armour or four inches (10cm) of oak and was said to be capable of pinning a man to his horse through both legs. Up to the development of firearms, the longbow was the outstanding weapon. Indeed, it could beat the musket for accuracy as late as the 1790s.

The crossbow was the hand version of a Roman siege weapon, the ballista, which shot javelins more than 300 yards (275m). There were several Roman siege engines, invented to break down the walls of forts or shoot objects over them. These objects might be rocks, incendiary darts, dead carcases to start plague, or even the occasional live prisoner.

The biggest siege engine was the onager. Like the ballista and the large catapult, it was built of wood,

Below: A medieval crossbow, which worked on a similar principle to the ballista.
Below right: This sinister weapon was known variously as a holy water sprinkler or King Henry VIII's walking staff. Made about 1540, it combined a three-barrelled gun, a spear, and a spike-studded mace.

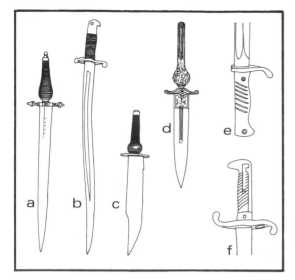

and the firing principle was the same in each case: a twisted skein suddenly released.

Similar weapons were used right through medieval times, the forerunners of artillery guns. Among them, the trebuchet would no doubt be classed as the heavy artillery. A whole tree was used as a pivot in this monster see-saw which, from a sling, threw boulders 200-300lb (90-136kg) in weight over 600 yards (550m).

About 1120 the Arabs invented the earliest known cannon, the Madfaa, a wooden bowl containing gunpowder with the cannon ball balanced on the rim. This early crude idea was adapted to create the more widespread Pot-de-Fer, an iron cannon shaped like a wine bottle, which fired arrow-like bolts.

It was not long before the early Pot-de-Fer was replaced by bigger cannon, perhaps as early as the Battle of Crécy in 1346. From then on, it was a case of slow improvement, in both the design of cannon and the explosives charge. The gunpowder meal was improved by coarser grinding and the extra air trapped between the larger particles produced a flash explosion. Now cannon ceased to be large and clumsy toys brought along to frighten the horses; they became artillery.

The invention of the slow-match, literally a smouldering wick, made possible the appearance of the first 'hand gonnes', the natural successors to blowpipes. Like blowpipes, early hand guns were basically just hollow tubes down which a projectile was sent.

Improvements were made until the familiar hand gun shape began to evolve. The priming pan was invented, a small pan of powder outside the firearm which could set off the main charge inside.

The next big advance was the invention of the matchlock. With this device a slow-match in a clamp attached to the weapon could move forward and touch the gunpowder without the marksman having to take his eye off the target.

The wheel-lock followed, invented in Nuremberg about 1520. A wheel on a spring was wound up and when the trigger was pulled the wheel spun against a piece of iron pyrites, striking sparks which ignited the powder.

A development from the wheel-lock was the Dutch or German snaphance, in which the trigger released a cock which struck a piece of flint against a steel plate to make the spark. A natural development was the safety-catch to prevent accidental firing.

After the Spanish musket of the 1500s, the next great hand-gun was the English military flintlock of the 1680s. The renowned 'Brown Bess' made its appearance and continued in general use for 160 years. It had no rifling – the spiral grooving inside a gun barrel which imparts spin to the bullet and keeps it on course – so the important thing was not marksmanship but volume of fire.

The clumsy powderhorn or pouch continued until a Swede, Gustavus Adolphi, invented a cartridge in 1585. This paper cylinder contained both powder and ball; the end was bitten off and the powder poured into the pan.

It was a Scots minister, the Rev. Alexander Forsyth, who invented percussion, the next great step. In the early 1800s, he discovered that fulminates ignited better if given a sharp blow than when lit by sparks. His 'scent bottle' percussion lock was revolutionary and based on a hammer released by a trigger.

Percussion made possible the invention of pistols and a wide variety of models followed. One of the early curiosities was the 'Pepperbox', with up to six barrels, each loaded separately at the muzzle.

What percussion did for pistols, rifling had already done for the bigger hand-guns in the 1700s. In the early days of cannon, it was found that the often impure gunpowder clogged barrels. So German gunmakers started to cut grooves on the inside of barrels to collect the dirt. It was only a step away from giving the grooves a twist and turning them into rifling.

Another important development was the invention of breech-loading. For hundreds of years, all guns were loaded by pushing the charge and the projectile down the barrel of the gun. Experiments in making a gun which would open at the other end to allow loading were made in the 1500s, but failed because the breech – the rear opening – could not be made to close sufficiently tightly.

The final step came in the 1860s when a French officer, Col. C. Ragon de Bange, devised the

Far left: Bayonets, first invented at Bayonne, France, in the early 1600s and modified many times since.
a, plug bayonet which fitted into a musket barrel;
b, Yataghan bayonet, London 1860;
c, Bowie plug bayonet;
d, fancy hunting bayonet;
e, Mauser butcher bayonet, German, 1914;
f, German, 1875.

Far left: James Puckle invented these two repeating guns in 1718 and 1720. His intention was that the guns should fire round bullets against Christians and square ones against Turks!
Left: This French catapult, one of several designed for use during World War One, was intended for hurling hand grenades at the enemy.

Right: These early models of the tank show clearly how the vehicle came to have its name – used at first as a security measure.
Far right: This German 'Goliath' beetle of World War Two might have been better called 'David' — it was a remote-controlled demolition vehicle designed for use against tanks or strongpoints.

Below: The US ram warship 'Dictator' of 1863, one of a number of coastal and river craft, sitting low in the water and unsuitable for ocean warfare.

Right: HMS 'Ark Royal', one of the first British aircraft-carriers to be purpose-built. She was employed as a seaplane-carrier in the Dardanelles campaign of 1915.
Far right: The German V-1 flying bomb of World War Two, photographed immediately after launching.

The 'bouncing bomb', invented by Barnes Wallis for the RAF to destroy the Mohne Dam in Germany during World War Two. This picture shows the bomb being dropped on its target.

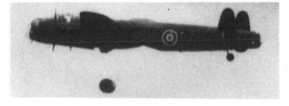

interrupted-screw-thread breech. It had been found that a long screw-thread gave the best breech closure, but took a long time to screw shut. By having the screw-thread cut into sections lengthways, de Bange made a breech that could be closed with one-twelfth of a turn. Simultaneously, Alfred Krupp in Germany evolved an equally satisfactory breech closed by a sliding wedge. Breech-loading rifles were developed between 1800 and 1850.

Most early cannon were made of brass or bronze. Cast-iron cannon came into use in the 1700s but, though they were stronger than bronze cannon, they were not strong enough for the improved explosives that were being developed. The answer was to build up the gun-barrel from several layers of cast and wrought iron, a process designed by an American engineer, Daniel Treadwell, in 1841 and developed by the British ordnance engineer William Armstrong a few years later.

Slowness in reloading was always the fault in early guns and even in the 1500s attempts were made to produce a rapid-fire weapon. The earliest usable

version was probably the Puckle gun, invented by an English lawyer, James Puckle, in 1718. It had one barrel and six or more cylinders turned by a cranked handle.

In the mid-1800s came the Gatling gun with multiple barrels. It had a hand-crank and could fire up to 800 rounds a minute. It was patented by the American Richard Jordan Gatling in 1862. A French version, the *Mitrailleuse,* could fire up to 125 rounds a minute.

The first true machine-gun was invented in 1889 by another American, Hiram Maxim, working in London. It was a single-barrel weapon, completely automatic, all the actions of loading, firing and ejecting empty shells being performed by its own recoil.

In World War Two, the light machine-gun became the chief infantry weapon of all the armies. The British Bren, invented in Czechoslovakia, and the German Spandau led in turn to the semi-automatic weapons, such as the American carbine, capable of both single shot and rapid fire, and later still to the sub-machine-gun or machine-pistol.

Developments in ammunition paralleled the developments of the guns that fired them. The earliest shot was round pieces of stone, soon to be superseded by balls of bronze, iron or lead. As an anti-personnel device came the invention in the early 1400s of case-shot: a canister filled with small projectiles fired from a big gun. In 1793 an English soldier, Lieut. Henry Shrapnel, devised an improved form of case-shot, with a fuse which split the canister open before it reached its target, so that the musket balls the projectile contained were spread over a wider area.

The shell, a canister containing explosives which bursts on its target, was first used in 1690 by a French sea captain named Deschiens. But it was not re-invented until 1788, when an English sailor in Russian service, Samuel Bentham, used shell with great effect against the Turks.

Gunpowder remained the only explosive until 1866, when the Swedish chemist Alfred Nobel invented dynamite, a form of nitro-glycerine which was safe to handle but twice as powerful as gunpowder. A British variant, cordite, was developed in 1890. TNT (trinitrotoluene), first invented in 1863, came into general use much later when detonators were developed. Detonators are used to set off relatively safe explosives, like TNT, and consist of substances such as fulminate of mercury which can be set off by percussion.

A development from prehistoric Man's throwing-stone was the hand-grenade, first used in the 1400s, then going out of use until the late 1700s and return-ing in a more sophisticated form by World War One.

World War Two brought about the invention of the two weapons which seemed to be the ultimate in weaponry: rockets and nuclear bombs. They were preceded by another weapon which had a short life: the flying bomb, called the V-1.

The flying bomb was a German invention. It was a small jet-propelled, unmanned aeroplane carrying a 2000lb (900kg) warhead and its controls were preset before launching to take it to its target area. But the flying bomb was not accurate and in time it became possible to take countermeasures which largely nullified its effect.

The first rockets were also a German invention, the work of a large team of scientists headed by Wernher von Braun, later renowned for his work on space flight. The rockets, known as V-2s, also carried 2000lb (900kg) warheads and they were launched from sites in western Europe to bombard London. These rockets had also limited accuracy, but they did a great deal of damage before their launching sites were captured.

Since World War Two, a great many rocket weapons have been developed from the V-2s, the main feature of which is accuracy in hitting the target. These guided missiles, as they are known, are made in a great many varieties and most of the research and invention has been the work of American and Russian engineers. But all the different types can be classified in four groups.

Surface-to-air missiles are launched from the ground or ships at aircraft or other missiles. Surface-to-surface missiles have surface targets, while air-to-surface missiles are launched from aircraft at ground objectives. Finally, air-to-air missiles are used in aerial combat.

All these missiles can be guided in one of several ways. Some are steered from the launching site by radio; some follow a radar beam which is directed at the target; some are preset, like the original V-2; and some 'home-in' on the target. The homing missiles are guided to their targets either by heat radiating from them – particularly used against aircraft – or by sound – used against submarines. A few locate the target by radar and then lock on to the beam.

The atomic bomb was also the work of many people and the final result of years of research and invention. It began when the Italian-born physicist Enrico Fermi, working in the United States, invented an apparatus which produced the first atomic chain reaction. That was in 1942, and work went on round the clock for $2\frac{1}{2}$ years until the first atomic bomb was test fired in New Mexico in 1945. Just 21 days later the second bomb was dropped on the Japanese city of Hiroshima, on August 6, 1945, and that was followed by a third bomb on Nagasaki three days afterwards.

The early atomic bombs, and many of those made since, depended on the splitting of atoms of heavy material into lighter atoms. The first fusion bomb was detonated by the US in 1951. It was activated by the fusion of atoms of hydrogen, the lightest chemical element, to form atoms of helium. Whereas the first atomic bombs produced a force equal to around 20,000 tons of TNT, hydrogen bombs have been detonated with a force equal to 65 million tons of TNT.

In addition to all the proven weapons, there have been many weird inventions which did not progress beyond the drawing-board and others which were centuries ahead of their time.

Leonardo da Vinci, who invented new workable cannon in his own day, left plans for a steam cannon and an 'organ' gun having multiple barrels, a chariot with whirling scythes and an armoured car, propelled by crankhandles, which could reasonably be described as the first tank design.

Another tank-like weapon was the remote control German 'beetle' of World War Two. Controlled by electric cables and limited to a range of 825 yards (755m), it had a charge of 183lb (83kg) and travelled at 5 mph (8kph). It was intended for attacking pillboxes and tanks.

Above: A Polaris ballistic missile being fired from a submarine during tests.
Left: A secret service weapon – the Soviet KGB's noiseless gas pistol, which shoots hydrogen-cyanide darts.

Far left: What that famous mushroom cloud grew from – the 'Little Boy' atomic bomb which destroyed Hiroshima, Japan, on August 6, 1945. It was detonated by radio 560m above the ground.
Left: A 'Wasp' helicopter carrying a torpedo.

Above: An FH-70 155mm gun-howitzer which can hurl a shell for about 30 km. It was developed in the early 1970s jointly by Britain, West Germany, and Italy.
Above centre: A supersonic Jaguar GR Mark 1 strike aircraft, developed particularly for battlefield support and low-level reconnaissance.
Above right: A 'Chieftain' combat tank, carrying a 120mm gun, on manoeuvres.
Below: The 'Mighty Atlas' inter-continental ballistic missile being test-fired.
Below right: A 'Phantom' jet strike plane carrying underwing air-to-air missiles, with more missiles and a Gatling cannon in the foreground.

Among the inventive Victorians, the Scottish engineer James Nasmyth came up with an idea for a 'floating mortar' which foreshadowed the later miniature submarines and human torpedoes. Nasmyth's model was tube-shaped, steam driven and had a large mortar shell sticking out of the front to be rammed underwater against the side of an enemy ship.

From the earliest days of guns, combination weapons were invented which incorporated pistols and trusty old weapons like axes, picks, daggers, halberds and double-tined forks. A grim sense of humour gave the name 'Holy-water Sprinkler' to a hand-mace of the 1500s having nine spikes around the weighted head, a frontal spike and three concealed gun barrels.

The most widespread combination weapon was the rifle and bayonet. When French peasants hunting bandits ran out of gunpowder, they improvised by ramming their hunting knives into the muzzles of the arquebuses. By 1650 the first specially designed blades to fit a gun had been invented in Bayonne.

With the development of pistols, a whole new world of combination and concealed weapons became possible. There was the pistol with a knuckle-duster formed out of its butt.

In 1814, John Day of Devon invented one of the most successful concealed weapons: the cane-gun, a popular means of defence against footpads, fierce dogs or poachers.

Whether for self-defence or secret service work, concealed weapons have continued to inspire the inventor. Guns have been combined with umbrellas, torches, pipes, watches, cigarette cases, vacuum flasks, cameras, helmets and workmen's tools. There was even the booby-trap sporran, which contained a metal purse fitted with a four-barrel flintlock activated by the purse knobs. In World War Two, the Nazis had a belt-buckle pistol which fired four shots in unison.

One of the earliest trap-guns was designed to fire when tripped by an intruder. Probably first used against poachers and grave-robbers around 1800, it consisted of a rotating gun, at the hub of several trip-wires, which would swing round to point at its victim.

One of the most widely-manufactured of the concealed guns in the 1890s was the palm pistol, round or square-shaped and about hand-size. It usually had a short barrel which pointed between the fingers and was fired by compressing the fist.

Designed for personal use, palm pistols had obvious advantages for secret and criminal use. It was a squeeze pistol, for instance, which the anarchist Leon F. Czolgosz used to assassinate America's President William McKinley in 1901.

Today's secret-service weapons include gases which immobilize the victim and an electronic gun that fires into the victim two darts which set up a lethal 50,000-volt electric circuit.

And a development of one of Man's earliest weapons, the poison-tip dart, is among the very latest secret service weapons. Snake poison or a shellfish toxin capable of producing instantaneous and untraceable death can be used on the darts. They are fired from silent electric guns or from such innocent-looking everyday objects as adapted fountain pens.

Potato thrower

One of the simplest yet most ingenious weapons of World War Two was invented in 1939 by a British engineer Treve Holman, whose great-grandfather helped Richard Trevithick make one of the first steam locomotives.

It was the Holman Projector, which started life as a compressed air gun, and consisted of a mild steel tube connected to a compressed air supply. If you dropped a projectile down the tube, it triggered a blast of air which hurled the missile high into the air.

In 1940 a version of the projector working off steam was fitted into small ships such as trawlers. The weapon soon became known as the 'potato-thrower', because friendly ships used it to engage in mock battles with 'spuds' as ammunition. Firing hand-grenades, 'potato-throwers' accounted for several hostile aircraft in real battles.

Sanitation

Far right: The hot bath in the ruins of a Roman villa, showing the underfloor heating system powered by hot air.

Opposite: The warm bath at a 19th-century Turkish bath in Germany, fed by a natural hot spring.

The history of baths and bathing is at least 5000 years old. At Mohenjo-Daro in the Indus River Valley of Pakistan, archaeologists uncovered a public bath nearly 1000 sq ft (93 sq m) in area, dating from about 3000 BC. Even the private houses there had their own bathrooms, fitted with terracotta pipes encased in brickwork. Efficient taps controlled the water flow.

A thousand years later, the royal families of the Minoan palace at Knossos in Crete luxuriated in 'modern' bathtubs that were filled and emptied by vertical stone pipes cemented at the joints. The Minoans had also invented glazed pottery pipes which slotted together very much like present-day ones.

Baths in ancient Egypt were often associated with religious ceremonies: priests took cold baths four times every day. But by the 1300s BC, Egyptian aristocrats and wealthy families, too, had their own well-appointed bathrooms. The Egyptians had invented copper pipes well before 2000 BC.

The religious aspects of bathing were carried to great lengths by the Jews under Mosaic law. With them, bodily cleanliness and moral purity were synonymous. Under the rule of David and Solomon, from about 1000 to 930 BC, complex waterworks were constructed in Palestine.

In ancient Greece, Homer's heroes bathed in tubs of polished wood or marble and enjoyed the luxury of hot water. However, from about 499 to 479 BC, the Spartans literally poured cold water over the 'flabby' idea of warm baths, and introduced the more manly cold shower. Mind you, in sunny Greece a cold shower is often very welcome, so perhaps the Spartans had a point.

A few years later, the Athenian authorities relaxed the rule of austerity and built large, warm, public baths. These became well-patronized social centres at the time of Socrates – the 400s BC – but the bathers were allowed to sample the warm water only after first enduring a cold plunge.

Public baths and bathing reached their acme of

development under the Romans, from about the second century BC onwards. As with the Greeks, the baths became meeting places for society. But the Romans, with their increasing love of luxury, added all kinds of extras, such as gardens, shops, libraries, exercise rooms and open spaces for poetry readings. The largest Roman baths could accommodate up to 3600 bathers at a time. At first, men and women bathed separately and at different times. But later, mixed bathing became the fashion, despite initial imperial disapproval.

There were definite rules of procedure for the Roman bather. These resembled the different stages of a Turkish bath today. After a limber-up in the exercise-room, the bather stripped, rubbed himself with oil and sand (soap had not yet been invented), then stepped in turn into a warm room, a hot room and a steam room. Finally he plunged into a cold bath, after which he anointed himself with oil once more. To clean himself, the bather scraped off the mixture of oil and sand with a curved metal or bone 'brush' called a strigil. Cosmetics and perfumes were sold at the baths.

The water was heated by a furnace under the raised floor. Hot air from there travelled into a chamber, the *hypocaustum*, from which it was channelled into tubes. Water was brought to the baths by means of immense aqueducts, themselves a triumph of Roman architecture. Some of them, 2000 years old, still stand.

For smaller pipes the Romans used lead, bending sheets of it around cylindrical formers and soldering the seams of the tubes thus formed. The familiar extruded lead pipe of today did not appear until the early 1800s. In 1790 an English metalworker, John Wilkinson, devised a method of casting lead pipes around a steel former. In 1820 T Burr, a plumber from Shrewsbury in England, invented a lead-extruding machine in which a hydraulic ram forced the lead through a former to make seamless pipes.

From the decline of the Roman Empire, when the barbarians destroyed most of the baths and aqueducts, until the later Middle Ages, baths and cleanliness in general were little known. The orthodox Christian view in those days was that the flesh should be mortified as much as possible. Bathing and washing came to be regarded as pandering to the flesh – hence sinful. That view prevailed over most of Europe. Royalty and the aristocracy were able to compromise –

A Roman aqueduct: the Pont du Gard at Nîmes in southern France.

HISTORY OF BATHING

Modesty in bathing has changed a great deal down the ages. In medieval times, *above*, sharing a bath was not thought so very out of the way; the bath itself was a wooden tub surrounded by drapes. In their romantic moments the Victorians glamorized the nude, as in Lord Leighton's idealized pseudo-Roman painting 'The Bath of Psyche', *far left*. But when it came to the practical side and sea bathing in their own times, the Victorians used bathing machines, *left*, to preserve the would-be swimmer from view until he or she had actually entered the water.

Before the days of separate bathrooms, the bath for most people — of all ages — was a metal tub, often placed cosily in front of an open fire in nursery or bedroom.

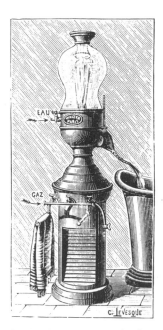

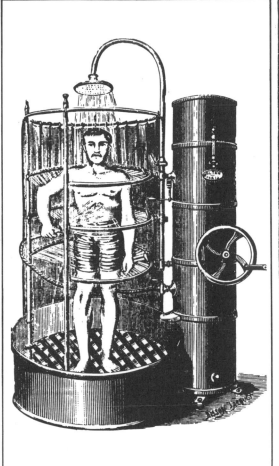

The Victorian era was a great time for inventions to help bathers. *Above,* from France came this 'instantaneous water heater' of 1889, which combined a supply of hot water, a radiator, and a towel warmer, all heated by gas; *right,* also from France was this pressure shower, which not only sprayed the bather from above and three sides, but also provided a hand-held jet to cover any parts that were missed; *far right,* a more sophisticated version, combined with a bath: it dates from 1882.

By the beginning of the 20th century the essentials of the bathroom were much as we know them today — basin, bath, WC and bidet, with plenty of mirrors. This one dates from 1905.

EOI–F

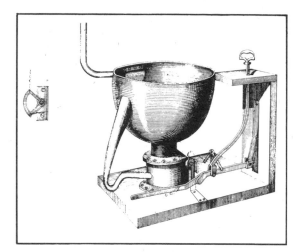

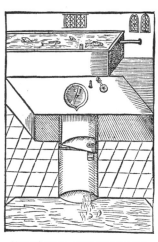

Above: A contemporary drawing of Sir John Harrington's pioneer water-closet. It discharged into a vault, which he advised should be emptied twice a day so that 'your worst privie may be as sweet as your best chamber'. For the same reason, presumably, he kept fish in the cistern!
Far left: Joseph Bramah's valve closet of 1778, a design which stayed in use into the late 19th century.

they bathed two or three times a year on special occasions.

Poorer people had little or no opportunity to clean themselves properly. But as the years went by, public baths began to appear in Europe, some of them supplied with hot water. By the 1400s they had earned a deserved reputation for promiscuity. Mixed bathing took place in large swimming pools, where music, food and drink were readily available.

Doctors of the time, as well as theologians, actually condemned bathing, because the practice was thought to spread disease. Judging by the state of public baths in those days, where hygiene was non-existent, they were probably right in their assumption. People applied strong perfumes and pomades to their bodies and hair, to disguise offensive odours.

Muslims, in contrast to European Christians, were fanatical in their cleanliness. During their occupation of Spain in the Middle Ages, the Moors built public and private baths wherever they settled.

But from 1400 onwards in Europe, the effects of the Reformation and its aftermath exacerbated the lack of hygiene, with Protestants and Roman Catholics outdoing each other in shunning pleasures of the flesh, including bathing and washing. Plumbing was negligible in western Europe and in North America. The coming of the Industrial Revolution in Britain did nothing to help, and industrial waste from the many factories piled up with the other refuse in urban slums. It was only after an outbreak of cholera had hit London in 1832 that the British authorities began a campaign of building public baths and wash houses.

From that time on, the British led the world in constructing public wash places and in devising and manufacturing elaborate plumbing with which to equip them. Sinks, baths, shower-baths and public swimming pools became comparatively common in Europe. Later, medicinal baths in the Roman style were built at places such as Bath in England and Spa in Belgium, where natural mineral springs flow out of the ground.

Towards the end of the 1900s the private bath, instead of being an incidental luxury item of furniture, was elevated in status to occupying a room of its own in better class homes, and the bathroom was born. Hot-water systems were expanded to cater for the new room. Most bathtubs were made of iron, copper or wood. Many of them were hooded or disguised to look like couches when not in use, because bathroom and other sanitation fixtures were regarded as vaguely improper.

The Scythians, from the Ukraine, are credited with the invention of the steam bath some 2500 years ago. Their bath consisted of glowing hot stones over which they poured water to raise the requisite steam. The whole apparatus was enclosed in a small tent, in which much of the steam was trapped.

The Finnish sauna is believed to have derived from the Scythians' steam tent, which may also have been the ancestor of Russian and Turkish baths incorporating a hot room, a steam room, a cold shower and an attendant who scrubs or massages the bather.

Sea bathing is a comparatively recent development. For hundreds of years, the oceans were regarded as obstacles to be negotiated or, in the case of familiar coastal waters, as convenient trade routes. It was not until the evolution of the bathing machine about 1790 that people generally took advantage of sea water for cleanliness and sport. The first bathing machines were used at Margate, in Kent. They were rolled down the beach into the shallow water so that the occupants, barely showing an ankle in their voluminous raiment, could sidle down the steps into the surf with maximum modesty.

Sewerage systems, drains and water supply were all highly developed thousands of years ago among the ancient Cretans, Egyptians and Incas. It was the ancient Greeks who first divorced disease from the magic that had surrounded it among primitive peoples and ascribed it to natural causes – an imbalance between Man and his environment. They recognized hygiene as an aid to healthy living, and organized public health services accordingly.

But it was left to the Romans, who inherited the Greek ideas on health and disease, to establish the best possible defences against filth and pollution. Apart from their many public baths and the water supply system they constructed, the Romans built a gigantic sewer, the *Cloaca maxima*, in the sixth century BC, to drain the site of the Forum. It still functions as part of Rome's drainage system. In the AD 300s, there were 150 public lavatories in Rome. Pure drinking water was carefully separated from water intended for washing purposes.

London had a complex main drainage system built by the 1200s, but nobody was allowed to let 'offensive waste' pass through it until 1815. Although Parisians could boast of a sewerage system by the 1400s, still only one Paris house in 20 was linked to it by the late 1800s.

One of the great problems exercising the minds of the authorities who governed the medieval towns was their very impregnability. They were virtually walled fortresses and, in making it difficult for unwanted people and material to get in, the town planners of the day made it equally difficult for things – including effluent of all kinds – to get out.

Townsfolk and animals shared the precious land enclosed by the walls and towers, and littered the unpaved roads with filth and garbage.

The cleanest places in those crowded communities were almost always the monasteries and the markets.

In the 19th century, a thing of utility could also be a thing of beauty, like this elaborate WC pedestal.

Above: Pumping engines like this were built for keeping the water supply up to pressure.
Above centre: What the eye seldom sees: major outfall sewers like this lie deep under the world's cities.
Above right: And what the eye did see: a typical public convenience in the 19th-century streets of Paris.

The monks built their settlements to well-thought-out plans, with efficient latrines, ventilation and water supplies. As for the markets, the medieval health authorities had a horror of rotting food, regarding it as a much likelier source of disease than the piles of effluent which littered the streets. Any waste food and scraps that might decompose were hastily swept from the market area.

That essential convenience of modern living, the watercloset, was probably invented about 1460, but the first person to do anything really practical with the idea seems to have been an Elizabethan courtier, Sir John Harrington. Banished from court for a while by his outraged monarch for translating and circulating among her ladies a racy story by the Italian poet Ariosto, Harrington retired to his home near Bath.

There, besides translating some more Ariosto, he worked on a design for a flush WC and duly installed the contrivance in his own home. Having won his way back into Elizabeth's good books, Harrington fitted a royal flush WC into the queen's palace at Richmond, Surrey. Unfortunately he wrote a book about his device, called *The Metamorphosis of Ajax* – the title is a pun on the old name 'a jakes' for a privy. The book appeared in 1596, and its earthy humour led the incensed Elizabeth to banish Harrington from court once more.

An improved model of Harrinton's WC, incorporating a stink trap, was patented by a London watchmaker, Alexander Cumming, in 1775. Further improvements were made by a London cabinet-maker and inventor, Joseph Bramah, in 1778. These early WCs were connected straight to cesspits and, even after the invention of stink traps, the smell from them must have been pretty powerful. Not until the invention of a modern sewer system in Hamburg in the 1840s, with arrangements for flushing the pipes regularly with river water, was the general health of people sensibly improved. The finest of the 19th-century sanitary engineers was Sir Joseph Bazalgette who, in the 1850s, equipped London with an efficient system of sewers for which he invented automatic flood doors and new pipe sections which allowed a speedier flow of effluent.

The chamber pot, from humble beginnings, offered scope for improvement. In Victorian times it became a veritable *objet d'art*, and even in the 1900s appealed to inventors as a vessel that might be elaborated. In 1929, for instance, an American electrician, Elbert Stallworth, patented the first electric chamber pot for use on chilly nights. In a rubber and asbestos seat which ran round the upper edge were embedded metal bands enclosing resistance wires between the mica strips.

As late as 1966, many inventors were still taking their thoughts to the WC. In that year, a Chicago hairdresser took out a patent for a novel toilet seat which embodied a buttock-stimulator for relieving constipation and for general massage. An electric motor set two separate halves of the seat moving backwards and forwards alternately. Both halves could also vibrate together at high speed.

On its way from reservoir to home, water passes through filter beds like this to make it pure and drinkable. The system was invented by a British engineer, James Simpson, in 1829.

Tear sheets

The inventor of toilet paper will probably never be identified but the invention of the perforated toilet roll is ascribed to an English manufacturer, Walter James Alcock, in the 1880s. Invention was one thing, but in the Victorian age marketing was another, and Alcock spent several years of hard selling before he was able to overcome the prudery of the general public and mass-produce his ingenious product.

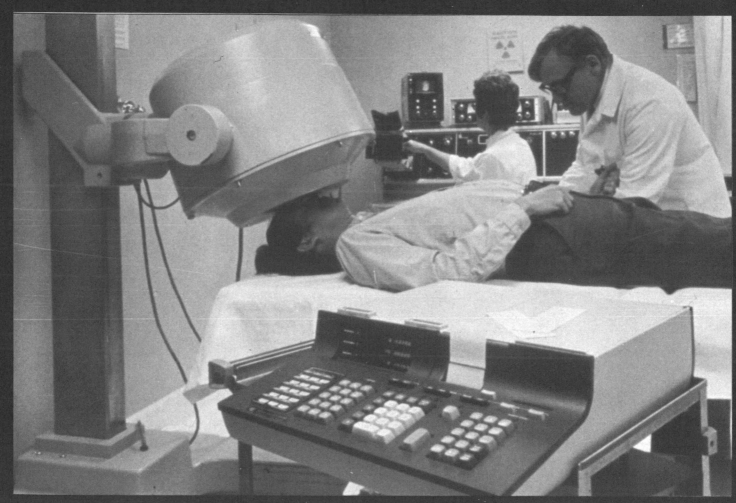

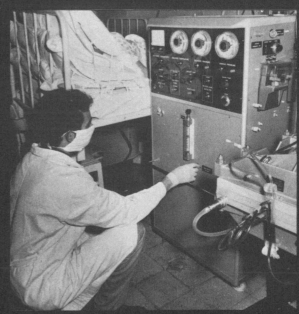

Ear trumpets have been used since the late 1600s. These two examples date from around 1800.

Opposite: Diagnosis by computer: a doctor uses the latest in electronics inventions to help in diagnosing a tumour.

An early compound microscope; this invaluable aid to medical practice was invented in the Netherlands around 1600.

Opposite: A patient receiving haemodialysis — treatment for the removal of impurities from the blood, with the aid of an older type of artificial kidney machine.

Man would be helpless without his senses and his limbs. When they are injured, he needs them to be repaired or replaced. Through the ages, inventors have responded to the challenge and done their best to imitate nature and repair the deficiencies.

Their efforts fall into two categories: physical aids to improve weakened organs and limbs, and artificial replacements for parts which have been injured beyond repair. The artificial part is called a prosthesis, and the branch of medicine involved is known as prosthetics.

Impaired sight has always presented a demanding challenge to inventors. As long ago as the 1200s, the English scientist Roger Bacon made a pair of rudimentary spectacles. In the 1700s, when the principles of refraction were first applied to the grinding of lenses, spectacles improved dramatically. Bifocal lenses were the invention of the 18th-century American statesman Benjamin Franklin.

Contact lenses of the scleral type – those that fit over the white of the eye – were first designed by Leonardo da Vinci in 1508. But the first such lenses were not actually made until 1887, when a German glassblower produced a pair. Early contact lenses of glass were so uncomfortable that people could not wear them for more than an hour at a time.

But in 1938 two Americans, F Muller and T Obrig, produced the first scleral lens of clear plastic which, being more flexible, gave greater comfort. The corneal lens, which covers only the cornea of the eye, was invented by the American optician Kevin Tuohy in 1948.

The ear trumpet for people who were hard of hearing first appeared between 1650 and 1700, and was a natural extension of the cupped hand. It was simply a long horn with a small opening for the ear and a larger opening for aiming at the source of sound.

In 1876 the American Alexander Graham Bell invented the world's first electronic hearing aid. It received sound pressure waves, converted them to a varying electric current, then turned them back again into sound pressure waves. In the 1920s, cumbersome and rather noisy hearing aids were made up of magnetic microphones and valve amplifiers powered by batteries. With the invention of the transistor in

Nasal comfort

So far as the nose and its comfort are concerned, perhaps the most useful invention has been the handkerchief. It was certainly known to the ancient Greeks and Romans – a dropped handkerchief was used as the signal to start the Roman games – and it may have been in use even earlier. During the Middle Ages it became a symbol of wealth, and people hung from their belts large showy handkerchiefs of cambric, linen or lace. Handkerchiefs were not carried in pockets until the 1700s, and children's handkerchiefs of that time were called 'muckminders'.

1948, hearing aids became miniaturized so that they could be hidden behind the ear or in spectacle frames.

The first surgical instrument was probably a sharp piece of flintstone. There is evidence to show that trepanning, cauterizing and circumcision were all practised by primitive Man, and the ancient Hindus used more than 120 different surgical instruments. Before 1842, alcohol and plant narcotics served as anaesthetics for patients undergoing surgery. In that year an American surgeon, Crawford Long, first successfully anaesthetized a patient with ether.

Trepanning – boring a hole in the skull – seems to be the oldest surgical operation known to Man. It was practised by the ancient Egyptians and by the early American indians.

The thermometer was invented by the Italian Galileo Galilei in 1593, but it was not a practical proposition for medical purposes until it was perfected by the German physicist Gabriel Fahrenheit in 1714. The compound microscope was invented about 1590 by the Dutch spectacle-maker Zacharias Janssen; but the first microscope used for medical research was a simple model made by the Dutchman Anton van Leeuwenhoek in the 1600s. The science of microbiology really developed after Leeuwenhoek saw what he described in 1674 as 'little animals in water' – protozoa and bacteria.

In 1816 a Breton physician, Réné Laennec, made the first stethoscope from a hollow wooden tube to which he attached an earpiece. When Laennec pressed the tube against a patient's chest, he was able to hear breathing and heart sounds. The first binaural stethoscope was created by an American pathologist and physician, Austin Flint, in the later 1800s.

The first porcelain-tipped probe is credited to a Frenchman, Auguste Nelation, also in the 1800s, and it was first used on the Italian patriot Guiseppe Garibaldi to trace a bullet. Doctors were able to administer drugs under the skin after the French physician, Gabriel Pravaz, invented the hypodermic syringe in 1853.

A significant step towards aseptic conditions in surgery was the introduction of sterile rubber gloves in 1885 by an American surgeon, William S Halsted.

In 1851, the German physicist Hermann von Helmholtz invented the ophthalmoscope, an instrument for examining the interior of the eye. His invention consisted of three thin plates of glass stuck together and mounted on a handle at an angle of 45 degrees. A light was placed by the side of the eye and the operator peered through the glass plates. Today's ophthalmoscope has lenses, a prism and a battery-powered electric light.

Around the end of the 1800s, when scientists were beginning to exploit a host of new discoveries, medical inventions proliferated. The sphygmo-

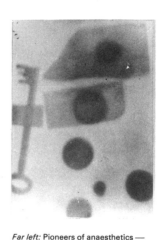

manometer for measuring blood pressure was invented in 1896 by an Italian, Scipione Riva Roci. It consists of an inflatable rubber cuff inflated by a ball, and a glass tube partly filled with mercury. The cuff is placed on the patient's arm and inflated. The physician listens to the pulse beat through a stethoscope on the arm, and the mercury in the tube indicates the amount of blood pressure on a graduated scale.

The high-voltage X-ray tube, invented in 1913 by an American physical chemist, William David Coolidge, provided a valuable medical aid.

The electrocardiograph was conceived when an English physiologist, Augustus Desiré Waller, discovered in 1879 a method of recording electrical impulses of the heart. The Dutch physiologist Willem Einthoven invented the string galvanometer in 1903, and the electrocardiograph was developed from this device. With each beat of the heart, a very slight electrical impulse develops and spreads through the heart muscle. The electrocardiograph detects and records these impulses, showing up abnormalities such as thickening of the heart walls, or impairment of the heart muscle. Einthoven was awarded the 1924 Nobel Prize for physiology for his investigations into electrical currents of the heart.

The sunlamp evolved gradually from research into the effects of sunlight on the human body about 1900. A Danish physician, Niels Finsen, discovered that tuberculosis of the skin can be treated by sunlight. For this advance, he won the Nobel Prize for physiology in 1903. Sunlamps give off invisible ultraviolet rays, which are the tanning rays in natural sunlight. These rays may be generated by a carbon arc, or a mercury arc inside a quartz tube, or from some other readily available source. Vitamin D is formed when the skin is exposed to such rays, whether natural or artificial.

The first practical encephalograph, a machine used to detect brain damage and disease, was developed in 1929 by a German, Hans Berger. It measures and records the voltages or electrical impulses given off by nerve cells in the brain.

The first fully practical fluoroscope was invented by Thomas Alva Edison in 1896. It reproduces on a fluorescent screen an X-ray outline of the patient's organs and bones. Several earlier but unsatisfactory models had been developed in England, Germany and Italy.

The first practical iron lung was developed in Boston in 1928 by two American doctors, Philip Drinker and Louis A Shaw. Used to treat paralysis of the muscles and organs of breathing, it is basically a large metal tank to which a pump is attached. The patient is completely enclosed except for his head, and the air within the tank is confined by a rubber collar around the patient's neck. As the pump extracts air from the tank, the patient's chest expands because of

the falling pressure, and he takes air in through his mouth and nose. As the tank fills with air again, the patient's lungs are deflated by the rising pressure. Some people have lived in iron lungs for years.

In 1953, an American surgeon, John H Gibbon, successfully used a heart-lung machine – or pump-oxygenator – in a major heart-operation. The machine's function is to perform during the operation the jobs normally carried out by the patient's heart and lungs. It takes venous blood from the heart, oxygenates it, then pumps it into the patient's leg artery. In this way, the body's cells are kept supplied with oxygen even though the heart may have stopped beating. With this apparatus, a surgeon can work for hours to repair or even replace the heart valves. Afterwards the heart is given an electric shock so set it beating again.

The first artificial kidney machine was developed by a Dutch inventor, Willem Johann Kolff, during World War Two. The kidney normally filters wastes, including urea, from the blood. If the kidney fails, too much urea accumulates in the blood, and poisoning results. By means of the artificial kidney – basically, a membrane and a pump – a continuously flowing stream of blood is diverted from the patient. On one side of the membrane is a salt bath. Waste substances from the blood pass through the membrane into the salt bath, and the cleansed blood is then returned to the patient.

Prosthetics, or 'spare-part surgery', has a history that dates from primitive Man's first attempt to plug the gap after losing a tooth. With the increasing violence of modern warfare and terrorism has come a demand for ever more ingenious and convenient artificial arms and legs. With the modern miracles of surgical implants and transplants, a great variety of spare parts can now be built into diseased and injured bodies.

Teeth are among the oldest 'spare parts'. Gold dental bridges, skilfully fashioned, have been uncovered in ancient Etruscan ruins, and it is believed that false teeth were worn by the Babylonians and ancient Egyptians.

Another replacement, perhaps more cosmetic than therapeutic, is the wig. Assyrians, Phoenicians, ancient Greeks and Romans all availed themselves of artificial

The electronic cardiac pacemaker has saved many lives; this model is worn externally, but many are implanted under the skin.

Above: This walking chair for invalids was a by-product of US space research into making an unmanned vehicle to carry scientific instruments on the Moon.
Below: Back in 1899 electricity was thought to be curative, as this advertisement for 'electropathic belts' shows.

hairpieces that disguised varying degress of baldness. The ancient Egyptians went to the trouble of shaving their heads and donning close-fitting wigs as a protection from the Sun's heat.

Spare parts for the heart became a reality in 1951 when an American surgeon, Charles Hufnagel, developed a plastic ball-valve to replace a faulty valve at the base of a patient's aorta. Hufnagel's device is a three-bar titanium cage enclosing a plastic ball, and the whole instrument is stitched to the heart muscle. The ball rises to each heartbeat and falls between beats.

Another cardiac innovation of Hufnagel's was the artificial artery of plastic, which he developed to help casualties of the Korean War. Today, artificial arteries are made of woven or knitted plastic yarn in many shapes, 5mm to 35mm in diameter. Seamless and unkinkable, they can be cut, clamped or stitched, and are almost indestructible.

The electronic cardiac pacemaker was introduced in the 1960s in the United States. In the normal healthy heart, a group of muscle cells triggers rhythmic contractions 60 to 80 times a minute. These contractions usually travel to the ventricles, where most of the pumping effort originates. When these contractions are blocked, as a result of disease, the ventricles contract at their own rather leisurely pace of 20 to 30 beats a minute. Usually, this rate of heartbeat is too slow to sustain life; at best, it results in lifelong invalidism.

The artificial pacemaker is a simple electronic device, housed in a plastic shell about the size of a small cigarette case. Driven by mercury batteries, it generates regular electric impulses through output leads whose electrodes are stitched to the heart. The whole unit can be conveniently attached under the ribs where it is handily located for occasional battery renewal. Its presence is betrayed only by very quiet clicks.

That method of attaching the pacemaker has a disadvantage, however. When the connecting leads are worn, another operation is required. A better system is to lead the connecting wires through a neck vein and thence into the heart, where they lie touching the muscle of the right ventricle.

Those people whose natural heart pacemaker functions intermittently do not need a continuously active artificial replacement; in fact such over-stimulation might put them at risk. For them a demand pacemaker has been developed. It works in the same way as the ordinary artificial pacemaker, but cuts in only when the natural pacemaker shows signs of flagging.

The demand for a complete mechanical heart has not gone unheeded by inventors, and various ingenious models and prototypes have been patented. But apart from the problem of rejection – which still impedes the success of natural heart transplants – there is also the problem of weight/strength ratio. The weight of an artificial heart of plastic and metal would put an intolerable strain on the connecting links inside the chest cavity. Surgeon-inventors are working to overcome this major impediment.

Eyes and ears have often been replaced purely cosmetically, but present-day experiments are aimed at bypassing the ear and eye. Scientists are exploring the idea of electrically stimulating certain areas of the brain to reproduce the sensations of seeing and hearing. Messages have already been successfully delivered and received by the auditory cortex directly, without involving the ear and its hearing mechanism.

In many ways, prosthetics has made greater if less spectacular advances in the matter of replacing legs and arms. Men were hobbling about on wooden legs as long ago as 500 BC. But it was not until during and after World War Two that new materials and techniques gave limb-fitters a fresh impetus.

In the 1790s, an amateur English mineralogist, William Gregor, discovered in the sands of Cornwall the chemically inert silvery metal called cobalt. It is cobalt and its various alloys which have largely displaced the steel, gold, silver, wooden, ivory and bone plates that were used before the war to bridge fractures; all these other materials posed problems of rejection.

In 1890, the first ivory hip-joint was used in Germany. Five years later, a British orthopaedic surgeon, Sir Robert Jones, improved on it with a joint made of gold foil. In the 1950s, two British surgeons, George McKee and John Charnley, introduced a femoral head made of metal that articulated with a cup made of plastic or metal.

Nowadays, most artificial legs are made of plastic or light wood. Joints at the knees and feet enable wearers to bend their legs at will.

Artificial arms and hands have been more difficult to construct. Usually the movement of an artificial arm is controlled by straps from the opposite shoulder. But progress is being made to control the hands directly by nerve impulses.

'Let Us Spray'

One of the greatest inventions in the history of medicine was a practice rather than an object: the introduction of antiseptic surgery. The man responsible was the English surgeon Joseph Lister, and he based his invention on the discovery, by the great French scientist Louis Pasteur, that organisms in the air caused wounds to turn septic.

In 1865, Lister introduced carbolic dressings for wounds, and used a carbolic spray in the operating theatre to try to purify the air – giving rise to a wry joke among his medical students, 'Let us spray'. The spray did not prove effective – but scrupulous cleanliness did, and thousands of lives were saved.

Time and Motion

Man's first timepiece was undoubtedly the Sun, which caused the shadow thrown by an upright stump to lengthen or shorten according to the time of day and the season of the year. That is how the first shadow-clock, or sundial, was born.

The ancient Greeks began to use shadow-clocks as long ago as 1450 BC, but the Chinese are believed to have predated them. In the Old Testament, sundials are mentioned in connection with Isaiah, dating from about 750 BC.

But sundials could tell the time only when the Sun was shining. Darkness and overcast conditions left Man guessing at the time. To overcome this problem, the clepsydra, or water-clock, was invented. Coming into use in ancient Egypt at about the same time as the sundial, it consisted of a pot full of water with a small hole in the bottom. The water leaked out at a known rate, and marks on the sides of the pot showed the level reached by the falling water at different times of night or day.

Later, this concept was reversed and people hit on the idea of filling a bowl gradually instead of emptying it. A bowl with a hole in it was placed on the water's surface. The water slowly leaked into it until, after a calculated period, the bowl sank. Such 'sinking bowls' are still used to tell the time in Algeria. The Chinese, too, used all kinds of complex water-clocks more than a thousand years ago. They divided their day into 100 parts.

About 2000 years ago, some anonymous inventor introduced the hourglass or sandglass. In this, two glass containers, each with a hole in the bottom, were joined, and sand flowed from one to the other in a known period of time. When the sand had finished flowing, the apparatus was upended and the whole process began again.

The ancient Greeks also used an instrument called a *hemispherium*. This was an ingenious device, based on the idea of a sundial, in which the shadow of a pin, rod, or plate fell across a curved path, which varied according to the time of year. The Chaldean astronomer Berosus produced a simplified version of this about 540 BC, calling it a *hemicycle*. This idea spread to most parts of Europe in the succeeding centuries.

A little later, fire was exploited to serve as a timekeeper. Notches were gouged in a lit candle, and the time it took to burn from notch to notch was measured. The same idea was applied to a length of burning string knotted at regular intervals.

The first mechanical clocks probably evolved in the great religious houses during the Dark Ages. The ringing of bells regularly summoned the faithful to worship and other observances at various times of day

and night. A labour-saving device was eventually thought up for ringing the bells automatically, and after that it would not have taken long to add a pointer and a dial to mark the passing hours.

The first clocks worked on a system of weights and toothed wheels. A weight, tied to a length of cord wrapped round a spool, was used to turn the spool – rather as a bucket, pulled by gravity down a well, turns the winding handle to which it is attached. The spool was linked through a system of toothed wheels to a pointer on a dial. But some means had to be found of controlling the speed of the weight as it descended, otherwise it would tend to run faster and faster. This problem was solved by the use of a device called an escapement, which allowed the driving force of the weight or weights to 'escape' little by little and at a controlled rate.

The first clocks, built about 1300, used what is known as a verge and foliot escapement. The verge could rock to and fro: it had two 'pallets' (flanges) which engaged alternately with the teeth of a vertical wheel, known as the crown wheel. The crown wheel was turned by the pull of the descending weight. As the verge rocked to and fro it allowed the wheel to rotate one tooth at a time, and so controlled the rate at which the weight descended — and the time keeping of the clock.

These early clocks were large structures of iron, and the weights were wound on to wooden drums. They were very much products of the blacksmith's art, though Giovanni de Dondi of Padua made one of brass and bronze in 1364.

These first mechanical clocks were not intended to tell the time. Their primary function was to show the movements of the Earth and planets, and to strike a bell at regular intervals. Their heavy weights and rather cumbersome mechanism meant that these clocks had to be hung on strong walls where there was plenty of room for them. As a result, the early clocks were hung in public bell-towers and monasteries.

One of the earliest weight-driven clocks was made by an Englishman, Peter Lightfoot, for the abbot of Glastonbury Abbey about 1335. Its working life lasted until 1835, when a new system of gears was fitted. The

oldest clock of which we have reliable knowledge was set up in the church of Beata Vergine in Milan, also in 1335. It had no dial or hands but struck an hourly bell. The oldest clock still in existence and working is in Salisbury Cathedral, in England. It was made in 1386 and, like the Milanese clock, has no dial. It ticks every four seconds.

There were disadvantages to the verge and foliot escapement worked by weights. Any imperfection in the teeth of the vertical wheel, or any dirt, would immediately cause a variation in the force exercised by the wheel, and the clicks would become irregular. This problem was not solved until Galileo was able to show, in the 1580s, that the swing of a pendulum took virtually the same amount of time regardless of the arc through which it swung, provided it was not too large. In 1657 Christiaan Huygens, a Dutch scientist, applied the pendulum to clockwork, and revolutionized the clockmaker's art by vastly increasing the clock's accuracy. Up to that time most clocks had one hand only, which indicated the hours. But after the pendulum was incorporated, accuracy was so much enhanced that minute-hands and in some cases even second-hands became the rule.

The next big step forward came with the invention of an improved escapement – the 'anchor' escapement. Traditionally, its invention has been credited to the English physicist Robert Hooke, in the early 1670s, but it is known that the London clockmaker William Clement was using the device in his clocks dated 1670. The escapement earned its name from its anchor-like shape, with pallets that resemble flukes. The teeth of the vertical wheel activated the pendulum. The teeth and pallets were so shaped that the wheel recoiled with any further swing of the pendulum once the wheel and pallet were engaged.

The increased accuracy resulting from the use of the pendulum and improved escapement meant that the makers could concentrate on clocks that would work for longer periods without rewinding. From 30 hours, which had been the limit, clocks were made to run first for a month, and eventually for as long as a year on one winding.

The heavier weights needed for these long-running clocks made it more difficult to attach them to the wall. The answer was to house the clocks in long, coffin-like cases, and that is how the 'grandfather clock' was born. At first, grandfather clocks were tall and narrow, but when it was realized that a long pendulum worked better than a short one, the cases became wider to accommodate the greater arc of swing. This, in turn, led to a whole new industry of specialized cabinet-making.

Nobody knows who first used a coiled spring instead of the usual weights, to drive clockwork, but spring-driven clocks first made their appearance in the 1470s. These clocks were compact and could be put in any position. As a result, domestic clocks lent themselves to fantastic designs and ornamentation. Moving figures, animal and human, danced and grimaced; ships rocked, flowers opened, and balls rolled to musical accompaniment.

But before these compact clocks could be perfected, there were two major problems to be overcome. One was the danger of a mainspring's snapping suddenly. The other was the fact that as a spring unwinds, so its force decreases. The first problem is still with us to some extent, but the second was solved by an ingenious device called a fusee. Jacob Zech, of Prague, is credited with its invention in 1525. But there is evidence to show that it was already known in the 1400s.

Two small drums stood side by side, joined by a winding cord or chain. One drum had straight sides and was driven by the spring. The other resembled a truncated cone with spiral grooves. When the spring first began to unwind, it turned the fusee by pulling the cord from the narrowest diameter, and the pull on the fusee was therefore least at that point. As the spring unwound, the cord was pulled from a progressively larger diameter. As a result, the final drive to the main gear wheels was almost uniform.

Small clocks that were driven by a mainspring had a balance wheel which swung back and forth at a steady rate by means of a balance spring or hair spring. One end of this was attached to the balance wheel and the other to the body of the clock. Robert Hooke originally experimented with the idea, using a straight spring. But Christiaan Huygens, using a delicate spiral spring, first adapted it successfully to clockwork in 1675.

Another form of escapement that improved time-keeping was the deadbeat escapement invented by George Graham in London in 1715. It worked on pendulum clocks by giving the pendulum an impulse near the centre of its swing, so that for the rest of its swing it was subject to very little friction. The escapement resembled the anchor, but its teeth and pallets were arranged differently so that the recoil was eliminated. Clocks with such an escapement were

Top left: A medieval clock in the market place of Venice. Like most clocks of the day, it had only one hand, showing the hours.
Above: This alarm clock was made in Italy in the 1400s, and was used to arouse the monks in a monastery for services.

Left: A French perpetual calendar of the late 18th century: it showed the day of the month, the year, the day of the week, and times of sunrise and sunset.
Above: Just in case the Sun's wrong — a combined sundial and clock.

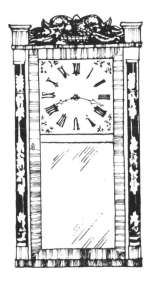

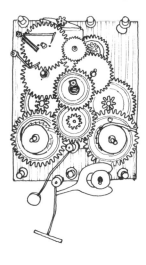

capable of keeping accurate time to within a few seconds a day.

Another problem that has always faced pendulum clocks is change of temperature. A metal pendulum will expand or contract if the temperature rises or falls significantly, altering the length of the pendulum and thus its time of swing. Errors of several seconds a day may be induced in this way. George Graham in 1721, and John Harrison in 1726, each came up with fairly effective temperature-compensation devices.

Graham used for the pendulum bob (the weight on the end of the pendulum that can be shifted up or down to adjust its rate of swing) a jar of mercury. When the temperature rose, the pendulum rod expanded downwards but the mercury expanded upwards, thus compensating for the increased length. Harrison invented the gridiron pendulum, made of brass and steel rods assembled so that when some expanded downwards, others expanded upwards, keeping the rod at a constant length. These rods were later superseded by zinc and steel tubes, but the principle remained the same. The problem was finally solved in the 1890s by the Swiss physicist Charles Edouard Guillaume, who discovered a nickel-steel alloy called invar which remained constant over a wide range of temperatures. As a result, invar was used not only for pendulums but also for balance wheels and springs.

Almost from their inception, mechanical clocks had embodied some form of striking mechanism. The striking train (the gears that drive the striking mechanism) is separate from the timekeeping part of the clock. At first, a driving weight was released each hour by the timekeeping mechanism and usually activated a hammer on a bell.

Until the 1600s the locking-plate striking system was used. This consisted of a wheel with a notched outer rim. A pallet falling into a different notch each time, as the wheel turned, determined the number of hits produced by the hammer. But in 1676, the English horologist Edward Barlow produced the rack, which involved a new kind of hour-hand with a stepped cam at its axis. The depth of the step opposite a pallet determined the number of hammer blows.

In the 1700s, spring-driven balance wheel clocks with hand-cut gears, of superior accuracy, began to appear. Because of their precision, they were given the name chronometers and became the standard timekeeping equipment of ships at sea, where pendulums would normally be useless. In 1761, the English clockmaker John Harrison invented a chronometer that varied less than two minutes during a rigorous five-month test at sea. An outstanding French clockmaker of the time, Achille Brocot, invented a pin pallet escapement for pendulum clocks, using agate or hardened steel pins. Another of his inventions was the Brocot suspension, in which the length of the pendulum could be adjusted from the dial front.

Alexander Bain, a Scottish clockmaker, invented the first electric clock in 1840. This was a longcase clock driven by electric current generated from coke and zinc plates buried in the ground.

In 1893, a German, Sigmund Riefler, invented a special escapement for observatory clocks. He devised a pendulum that had a nickel-steel rod supported by a short aluminium tube. He also invented a lever escapement for watches, in which the lever was fitted to the end of the balance spring.

In 1906, the Americans T B Powers and G and H Kutnow invented an electric battery clock, controlled by a large visible balance-wheel instead of a pendulum. At about the same time, the luminous dial was adopted in the US, using a mixture of phosphorus and radium with which to coat the hands and hour dots.

The first watches were made at Nuremberg, in Germany, about 1500. The oldest watch still in existence was made by a German locksmith, Peter Henlein, in 1504. It was really a portable, spring-driven clock. It was obviously impossible to use a pendulum in a watch, so the advent of the watch had to await the invention of the mainspring for motive power.

In the early 1600s the first watches protected by glass faces appeared. In 1704 the Swiss watchmakers Nicholas Faccio de Duiller and Peter Jacob Debaufe invented pivot-holes made of sapphires, for longer life, and the first 'jewelled' watches came into being.

In 1755 a superior escapement mechanism, known as the 'English lever', was designed by Thomas Mudge. The first pocket chronometer was produced in 1766, with a balance wheel of brass and steel. In 1790 the first wristwatch proper was made in Geneva by Jaquet Droz and Paul Leschot. J W Benson introduced wristwatches into England in 1885, but they were regarded as effeminate until their usefulness was proved in World War One.

In the meantime, the Americans were fully exploiting the possibilities of the watch. In 1878, D A A Buck invented the Waterbury watch, named after the place of its manufacture in Connecticut. This watch was a machine-made model in which the movement, together with the hour-hand, rotated in the case once an hour. It had 58 parts and a 10-foot (3m) mainspring.

In 1924, the English watchmaker and jeweller John Harwood patented the first self-winding wristwatch. There were no winding buttons, and the watch was wound automatically by movements of the wearer's arm.

The first electric watch was designed by Harwood in 1930, but production was delayed until after World War Two. In 1957 the Hamilton Watch Company introduced it into the United States. This electric watch worked off a small battery. The American Bulova Company was the first manufacturer to produce an electronic watch, some four years later. Called the Accutron, it embodied a mercury cell which vibrated a tiny tuning-fork. With only 12 moving parts and its fork vibrating 360 times a second, the watch was accurate to within two seconds a day.

In the 1970s watches as well as clocks began to use the oscillations of quartz crystal to bring timekeeping errors down to the level of less than one second a month – a standard of accuracy for ordinary watches which was undreamed of a few years previously.

Diving is as old as swimming. Homer's *Iliad,* probably written about 800 BC, refers to men diving for oysters as an everyday occurrence, so rudimentary underwater exploration must have developed much earlier.

The first divers merely held their breath and plunged, as Japanese pearl-divers still do. The first aid to diving was the diving bell, which had already been invented by the time of Alexander the Great in the 300s BC. Diving bells rely on their weight to sink to the bottom of the sea, and early models went down with just the air they contained, which was considerably compressed by water pressure, and became foul after it had been breathed for a few minutes.

The first person to try to pipe air down to a diving bell was the English astronomer Edmund Halley, who spent four years between 1698 and 1702 surveying the coast of the English Channel. His work was taken a stage further by his fellow-countryman, the engineer John Smeaton, who made improvements to the air pump in 1752, and fitted one of his improved pumps to a diving bell.

The diving suit evolved out of the diving bell, whose main disadvantage was that the men inside it could not readily move around. So inventors began to make smaller diving bells tailored to individual divers. The first true diving suit was invented by a German engineer, Augustus Siebe, in 1837. Modern diving suits are based on Siebe's design, which consisted basically of a heavy helmet, a waterproof canvas suit, and lead-weighted boots to keep the diver's feet firmly on the sea bed. An airline, a lifeline and a telephone cable connect the diver to the air pump and its crew.

Deep-sea divers in their clumsy suits can work only with difficulty, and they are tethered to their lifelines, which also hampers mobility. An obvious

Left: A naval frogman wearing a wetsuit, and with the latest self-contained breathing apparatus, returning from a dive.
Above: A conventional diving suit of about 1920.

development was for the diver to take a supply of air with him, leaving him free of all lines. A primitive attempt to free the diver in this way was made in 1825, but it was not until after World War One that really practical self-contained underwater breathing apparatus – scuba equipment – became available. The British firm of Siebe Gorman developed an oxygen apparatus, called 'Amphibian Mark I', during the 1930s, and similar apparatus was made for the German and Italian navies.

However, oxygen can prove poisonous at depths in excess of 30 feet (9m), and during the 1930s many experiments were going on with underwater use of compressed air. The modern aqualung apparatus, which permits dives of more than 100 feet (30m), was invented by two Frenchmen, Jacques-Yves Cousteau and Emile Gagnan, in 1943. They proved that light work could be performed at up to 210 feet (64m) below the surface, but that at depths below 130 feet (40m) nitrogen in the bloodstream causes sensations of drunkenness.

Special submersible vessels – in effect, totally enclosed diving bells – have been constructed for really deep underwater exploration. In 1930, an American zoologist, William Beebe, and an engineer, Otis Barton, invented the bathysphere – virtually a hollow ball attached to strong steel cable. Had the cable parted, the sphere would have sunk into the depths without a hope of rescue.

An improved submersible was the bathyscaphe, invented by the Swiss scientist Auguste Piccard just before World War Two, but not built until 1946. This craft was buoyed up by tanks containing petroleum, which is lighter than water, and ballasted by iron shot held secure by electromagnets. If the power supply failed, the ballast was automatically jettisoned and the craft surfaced. In 1960 a bathyscaphe made the world's record dive to a depth of 35,800 feet (10,900m) in the Mariana Trench of the Pacific Ocean.

Current research is aimed at creating underwater living quarters, where scientists may spend several days or weeks on the seabed examining ocean life.

Above: The first diving helmet, invented by Augustus Siebe in 1837.
Far left: An early 19th-century diving suit, designed for operation at comparatively shallow depths.

Below: A diving suit of the 1970s, with a magnesium alloy pressure body, articulated fluid-supported joints, and mechanical 'hands' controlled by the actual hands of the diver inside the suit.

Right: A frogman and a small seabed exploration vessel prepare to dive.
Far right, top: A model of the Piccard bathyscaphe Trieste, showing the crew chamber suspended below the huge flotation tanks.
Far right, bottom: Consub — short for Continental Shelf Submersible — a remote-controlled seabed exploration vehicle, designed to operate to a depth of 600 m.

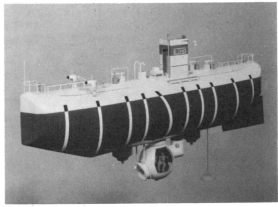

Above: Lowering a modern diving bell into the sea, part of an underwater oil drilling project.
Right: A VOL L1 submersible and HMCS Ojibwa preparing for a sea-floor rendezvous, during which men transferred from the submersible through the 'skirt' under the vessel to the Canadian submarine.

Business Machines

Before the twentieth century, most office jobs were done by hordes of clerks, book-keepers and secretaries, working laboriously with the simplest of tools. Then industrialization led to a vast expansion in the world of trade and commerce, with new products to be sold and new markets to be exploited. Machines had to be invented to help carry the load.

Many office machines appeared before 1900, but it was not until after that date that much of the equipment we know today was produced in its first really practical form. The 1950s saw further great advances, as the need to cope with escalating business administration led to the invention of entirely new systems based on automation, data processing and computers.

Over the years, the hardest-working office machine has probably been the typewriter, which was the apparently unattainable goal of inventors for many years before it finally made the production line.

One of the earliest attempts to invent a writing machine was made by an Englishman, James Ranson, in 1711. He is said to have designed a machine with keys like a harpsichord, and an inked ribbon.

The next mention of such a writing machine was three years later when England's Queen Anne granted a patent to Henry Mill for an 'artificial machine . . . for impressing or transcribing of letters'. However, no drawings or models survived, and we do not know if his idea was ever developed.

Several European inventors were working on similar ideas in the years which followed, but the first workable writing machine was produced by an American, William Austin Burt, in 1830. His machine, the *Typographer*, incorporated type mounted on curved bars on a wooden frame. A wheel turned the type to a position from which it was depressed on to the paper by a small lever.

But Burt's machine was too slow in operation to be practical, a fault which was to frustrate inventors for many years. Among the many writing machines which appeared briefly was that of Xavier Progin, who obtained a French patent in 1833 for his *Ktypographique*. With this machine he invented the principle of the typebar which is still found in modern typewriters.

Charles Thurber, an American, whose first typewriting machine appeared in 1843, based his designs on a circular index and a plunger for each character. He was the first inventor to use a cylindrical platen.

One of the few machines of this time to be widely used was the invention of a Mr Hughes of the

Henshaw Institute for the Blind in Manchester, England. He made a machine to emboss characters so that blind people could communicate with each other, and by 1851 his practical little *Typograph* was widely used in schools for the blind.

About the same time, another machine for the blind which both embossed and typed characters was the *Clavier Imprimeur* of Pierre Foucault, a blind French student. His machine was large, more like a table-top clavier, and the operator had to stand up.

A machine designed by an Italian lawyer, Guiseppe Ravizza, in 1858, with a wooden box-casing and a two-bank keyboard, looked more like the modern typewriter. His 1867 model had an interesting little device: when the end of a line was reached, a small hatch opened, a bell sounded and a sign flipped out with the words, 'The line has finished'. Ravizza was one of the first inventors to incorporate a moving inked ribbon.

Pastor Hansen of Denmark produced a *Writing Ball* about 1865 which foreshadowed today's 'golf ball' machines. Hansen was initally inspired by the desire to help the deaf and dumb to communicate. He designed a porcelain ball of block letters with plungers on top which looked not unlike a pincushion. The ball was placed above a half-cylinder which held the paper.

What is generally regarded as the first commercially successful typewriter was the American Sholes-Glidden

machine of around 1867, the first device actually to be named a *typewriter*.

Christopher Latham Sholes was a Milwaukee printer, editor and politician who arrived late on the office machinery scene: there had been some 50 typewriter inventors before him. His first machine, designed with the help of Carlos Glidden and Samuel W Soule, would print only one letter, but the three men felt they were on the right lines and they persevered.

Eventually, with the financial help of James Densmore and the practical advice of a mechanic called Yost, they produced a machine remarkably like today's typewriters. It interested Philo Remington, a manufacturer of guns and sewing-machines, who agreed to try it out. The first 1000 Remington machines came off the production line in 1874.

Within 30 years there were 30 typewriter-manufacturers in America. The business world was revolutionized, copperplate handwriting was outmoded, and career opportunities for women mushroomed.

Typewriters soon settled more or less into the design we know today, with front-strike machines, four rows of keys and the 'qwerty' keyboard.

Two of the most striking typewriter innovations of recent times were the 1941 breakthrough with proportional spacing, giving typewritten characters the appearance of printing, and the 1961 'golf ball' with type on a spherical shape which can be changed in a moment for another ball carrying a different typeface.

A machine developed in the early 1900s was the typewriter accounting machine which incorporated additional registers for adding and subtracting.

Later refinements allowed vertical and horizontal adding and subtracting, and then an accumulating mechanism took over most of the mental work. Electric machines were to follow, which computed automatic balances and totals; then came models which provided full typewriter and accounting keyboards in the one machine.

But these are recent refinements in the story of calculators, which goes back a very long way. One early method was the tally, a stick in which notches were cut to signify figures or quantities. When the tally was completed, it was split longitudinally, each party to a transaction taking half. At payment time the two halves were matched up, and if they tallied, the account was correct.

Tallies were used by the British Exchequer up to 1826. Eight years later, vast stocks of these old tallies were burned in the stoves heating the Houses of Parliament. They burned so well that they set fire to the whole place and destroyed it.

One of the earliest mechanical devices for calculating, and one still widely used today, was the abacus, a frame carrying parallel rods on which beads or counters are strung. Herodotus, the Greek historian who lived in the 400s BC, mentions the use of the abacus in Egypt. Other early users were the Chinese, Greeks and Romans.

In 1617, John Napier of Edinburgh invented Napier's Rods, marked pieces of ivory for multiples of numbers. In the middle of the same century, Blaise Pascal of France produced a simple mechanism for adding and subtracting. The English diplomat Sir Samuel Morland presented to Charles II an adding and subtracting machine he had designed.

Multiplication by repeated addition was a feature of a stepped drum or wheel machine of 1694 invented by the Leipzig philosopher Baron Gottfried von Leibnitz. This machine provided the basis for later developments. The first of these to be at all successful was produced by Charles Xavier Thomas in 1820. It was a calculating machine which would add, subtract, divide and multiply.

Ten years later, the Englishman Charles Babbage invented Babbage's Calculating Engine, a simple adding machine based on 'differences', but left it unfinished. Babbage is mainly notable for introducing the ideas of 'programming' and 'taping', and his ideas were later used in the Harvard Mark II computer of 1948.

These early calculators were largely impractical, but Thomas de Colmar's arithmometer of 1850, which worked by handle, was quick and accurate.

The first of the 'comptometers', a multiple-order, key-driven calculator, was patented in 1887 in the United States by Dorr Eugene Felt. It was followed by the William Seward Burroughs key-driven machine, based on the decimal system and able to deal with money columns.

A number of variants of the Burroughs design paved the way for modern machines. One of the notable developments was the mechanically operated differential analyzer produced by the electrical engineer Vannevar Bush at the Massachusetts Institute of Technology in 1930.

Up to World War Two, there were always some problems which could not be worked out in a short time on mechanical calculators. Then came the development of computers, initially in connection with anti-aircraft defence and gunfire control. With computers it became possible to answer in one second problems which would take mathematicians weeks to work out. The first digital computer was invented by Howard Aiken of Harvard University in 1944.

The development of electronics made it possible to reduce the size of computers so that their use could be more widespread, and by 1962 the first of the desk-top models for office use had appeared. It had no moving

Far left: This magnetic card typewriter records material as it is typed, and can then reproduce it as many times as desired, or act as a teleprinter for long-distance transmission.

The 'golf ball' typewriter, invented in 1961, can be used as a typesetter. The actual 'golf ball', *below,* can be changed in seconds to give a choice of more than 125 typefaces, while the machine itself possesses a built-in electronic memory.

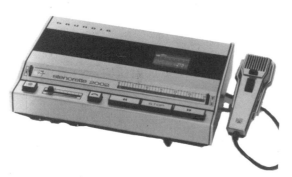

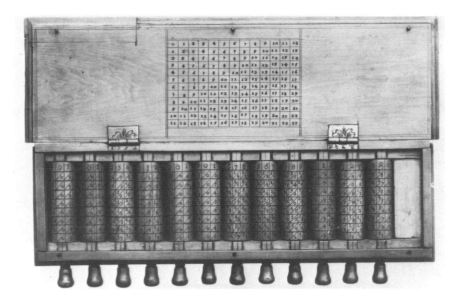

Napier's Rods, a calculating device invented by the Scottish mathematician John Napier, who also invented logarithms.

a machine which would produce five copies. James Watt of steam-engine fame, invented a machine to reproduce letters and drawings, and in 1803 Ralph Wedgewood patented carbon-coated paper in England.

A machine using a gelatin process and roller appeared in 1880. Known as the hectograph, it was invented by Alexander Shapiro in Germany. A few years later, A B Dick built the first mimeograph in the United States. This device used a stencil; the impression was placed in a set frame, and ink was applied to the stencil to make a copy.

Most duplicators today employ one of four processes. The spirit or liquid process involves aniline dye and solvent, a master copy and hectographic carbon sheets. The stencil process is much the same as the original invention. At first a stylus was used to make the stencil, then a typewriter was substituted. Now, originals can be produced by letterpress, photography, electronic scanning, teleprinting or brush and ink. A modern automatic stencil duplicator can produce 9600 copies an hour.

Two other methods of duplicating are *relief*, using a letterpress printing duplicator for which type is set, and *offset litho*.

Photocopying machines are a comparatively recent

parts, was completely silent and would do any sort of arithmetical calculation rapidly and accurately.

In addition, there are now pocket calculators based on integrated circuits which allow thousands of components to be reduced to a tiny silicon chip. They vary from models which simply add, subtract, divide and multiply, to sophisticated calculators capable of being programmed like computers.

For office work, there are electronic invoicing machines which will do 166 calculations a second. There are complete electronic accounting systems incorporating automatically-programmed accounting, typewriting and punch-card devices.

The expansion of office work quickly led to another prime need – the duplication of correspondence, reports and records. Early attempts to provide the means of duplicating included a device invented by the English physician, surveyor, and economist Sir William Petty in 1647. It consisted of two pens writing simultaneously.

James Young, a Scots solicitor, is said to have made

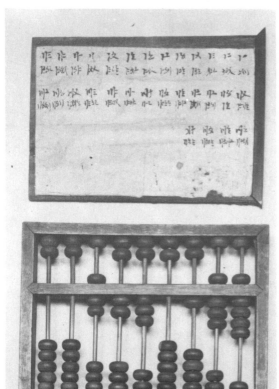

invention. They produce copies of the original simply by placing it in contact with a sentitized medium. There are infra-red models which produce copies in four seconds, and copier-duplicators which produce copies of an original at the rate of 40 a minute.

Microfilming is another means of copying which finds favour in modern offices, particularly where space has to be saved, as with newspapers and technical drawings. For instance, 8500 letters or 18,000 cheques can be micro-copied on 100 feet (30 m) of film.

Once it was possible to copy papers, it became necessary to file them. Up to the end of the 1800s this was done by pigeonholing documents in wooden compartments or spearing papers on wall and desk spikes.

Alphabetical divisions for easy reference came with the introduction of the box file and concertina file. About 1892 came the first vertical files, incorporating individual pockets. Then came filing cabinets as we now know them, with pockets suspended on rails and easily interchanged.

One of the later refinements for really big systems is a rotary system of lateral filing with pushbutton operation which brings selected folders in front of the operator.

Shorthand goes back to the ancient Egyptians, Persians, Greeks and Hebrews, who all used some system of signs to represent whole words or phrases. The first known system was invented by Marcus Tullius Tiro, Roman secretary to the orator Cicero, in the first century BC.

The Tironian system was used until the ninth century AD. Shorthand declined in the Middle Ages but reappeared in Elizabethan England when Dr Timothy Bright published a system which foreshadowed modern forms in 1588.

The first shorthand alphabet was described by John Willis in his 17th-century book, 'The Art of Stenographie'. A number of shorthand methods were invented in the next 200 years. One of the most popular was that devised by Thomas Shelton in 1626, which Samuel Pepys used for his diary.

Sir Isaac Pitman's 'Stenographic Soundhand', published in 1837, defined 40 sounds represented by lines and curves, and many abbreviations and contractions. It is still widely used. In 1888, an Irish-American, John Robert Gregg, published in London another system, also still popular, based on the cursive style of normal longhand.

In the late 1800s, machines were introduced which type a form of shorthand. They are known as stenotype or palantype machines.

Another device which supersedes shorthand is the

dictating machine. It was Edison's talking machine of 1878 that led the way to a graphaphone, patented in 1886 by Charles Tainter and Alexander and C A Bell, which would both record and reproduce.

Modern dictating machines use tapes, magnetic belts or discs. Refinements include an attachment for the typewriter which plays at the speed at which the operator types, and a battery-powered electronic pocket dictating machine.

Another development from the dictating machine is the Ansafone which automatically answers incoming telephone calls and records messages out of normal

Above: The first electrical tabulator invented by Dr Herman Hollerith of the US in the 1880s. Working with punched cards, it recorded the card data by moving pointers on the dials.
Below left: The first Braille writing machine was produced in 1892 by Frank Hall in Illinois, and helped blind people to work in business. This Perkins Brailler is a later development.
Below: A 7-bank printing and listing Hollerith machine of 1932.

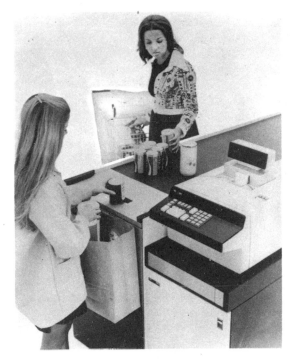

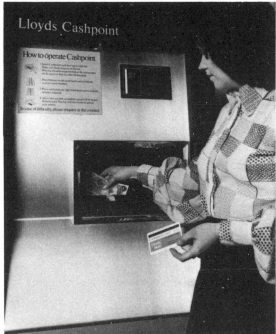

Right: Using an automatic electronic checkout system at a supermarket.
Far right: Electronic cash, too — one of the automatic money-dispensers now in use in banks throughout the world.

A lesson from the bees: this honeycomb storage bank holds hundreds of data cartridges which can be retrieved and brought into action at the push of a button.

working hours. It will also record conversations between two people or act as a dictating machine.

Modern offices have many convenience appliances from the paper clip, invented in Norway in 1899, to elaborate punched-card systems for sifting, sorting and counting records, particularly accounts.

Standard-sized cards for insurance records were known as early as 1855, but it was Babbage with his Difference Machine who introduced punched cards. In Babbage's system, one hole equalled 1, two holes equalled 2, and so on. With modern systems, it is the position of the hole which decides the number it represents.

When clerks did the sorting for the American 1880 census, the work took seven years to complete.

Punched cards were used for the census in 1890 and have been used ever since. They can do the work in a matter of hours. Today, punched cards are also used for programming computers.

Another widely-used appliance, the addressing machine, was first invented in 1870 when James McFatrich of Illinois patented the McFatrich Mailer. In such early machines, the name and address was cut with needle point type on to paper stencils. Modern machines, which can print 13,000 addresses an hour, use plastic cards, stencils in cardboard frames or embossed metal plates.

Letter-opening machines which can deal with 600 envelopes a minute, signature machines which sign 300 letters in five minutes, machines for dating, numbering, folding, sealing and franking are all part of the modern office scene: a long way from quill pens and parchment.

Sticky problem

No office is complete without one or more erasers – but without the tireless patience of one man, your eraser would be sticky in hot weather and brittle on cold days. His name: Charles Goodyear.

In 1834, Goodyear, son of a bankrupt hardware manufacturer, invented an improved valve for a rubber lifebuoy – and was told his invention was useless because the rubber perished too easily. Goodyear determined to make a better rubber. He spent five years experimenting, often in debt, ill from overwork and lack of good food, but never letting up.

Then success came in 1839 as a result of an accident. Goodyear dropped a mixture of rubber, sulphur and white lead on to a hot stove, and found the rubber became tough, just as he wanted it. He had invented vulcanization. It took him five years more to perfect the process. In the end Goodyear made comparatively little money from his invention, though it enriched others, and he died deeply in debt once more.

Industry

Far right: Industry depends on power, and this French watermill illustrates a form of power that has been used since about 3000 BC.
Right, lower: Another source of power that was used from ancient times until the 19th century, wherever there was plenty of convict or slave labour, was the treadmill.

All the great developments of modern technology are the result of Man's use of machines to do work for him. And all machines are basically developed from three simple ones: the lever, the wheel and axle, and the inclined plane.

The lever was known in ancient Egypt about 5000 years ago, and was probably old then. The wheel also existed about 5000 years ago; primitive wheels of about that date have been found in graves in Mesopotamia. But the principle of the roller, for moving large blocks of stone, was known at a much earlier date.

The inclined plane was undoubtedly known in ancient Egypt, where it was used in the form of ramps for hauling up stones to construct the Pyramids, also around 5000 years ago.

Industry, as we known it today, is not nearly so old. It began when men first started living together in cities and dividing their work in a specialized way so that each man did one thing and exchanged the fruits of his labour with others. Work persisted on an individual basis right up to the 1700s. The factory, as a place where plant and machinery are installed for making things, was an invention of the Industrial Revolution.

The success of the Industrial Revolution was due to power. Man at last conquered the problem of how to make something else do the work of his muscles. In a limited way, Man has been using power of one sort or another for many thousands of years. The first and most important form of power was undoubtedly fire, which releases energy in the form of heat. Peking Man was a fire-user at least 500,000 years ago. Archaeologists have found the remains of fires in the caves where he lived.

The first fires were started by natural causes, principally lightning, and for a very long time people had to keep fires burning constantly because once they went out, no-one could start them again. So far as we know, the first human beings to make fire for themselves were people of the New Stone Age, about 8000 years ago. But though fire was used for warmth and cooking, it played little part in their work for a long time. The first step in the development of other power sources was the domestication of the ox, around 6000 years ago, and with it came the first

invention on the history of power: harness. Yokes used to harness two oxen to a plough, have been identified from remains dating from around 4000 BC found in south-western Asia. The horse was domesticated about 5000 years ago, and the bit and bridle seem to have been invented about the same time. But yoke harness does not work so well on horses as on oxen, and it was not until the invention of the padded horsecollar in China about AD 500 that the full power of a horse could be used for anything except riding.

The sail was invented and used by early Man on his primitive log or bark canoes so long ago that it is impossible to say when. The windmill — the first use of wind to provide power on land — was a comparatively late invention, made in Persia some time around AD 600.

Water power was harnessed for driving mills long before the wind was used. The earliest watermill seems to have been invented around 3000 BC, in the hilly parts of south-western Asia, where swift-flowing streams abound. The watermill was unknown in ancient times in Egypt, where the Nile is for the most part slow-running.

The power generated by the waterwheel and the windmill is about 10 hp, and it was with power sources of this kind that the Industrial Revolution was launched in Britain in the mid-1700s. But it was the development of stream power which made that revolution so violent, and so complete.

Roughnecks old and new: *below,* drilling for oil in the pioneer days in Canada depended on manual labour: the men on the small platform at the left threw their weight back and forward, causing the spring pole to drop and rise and drive the drill bit deeper into the ground; *opposite:* workers connect a new length of drillpipe to a power drill aboard a 1970s ocean drilling rig.

Fuel and Power

Left: Col. Edwin Drake (with the beard) in front of his pioneer oilwell at Tinsville, Pennsylvania.
Above: Drilling for oil in the Caucasian field of Russia around 1900, using a percussion bit.

Wood was the earliest fuel available to Man, and at first he burned it as he found it, lying about the forest floor. It was probably observation of the way in which a partially charred branch from a former fire glowed and gave out heat which sparked off the first of a long series of inventions in the use of fuel: charcoal. Charcoal was the first fuel used to heat and work copper and bronze, a development that took place about 5000 years ago. It was also used for smelting iron, which began about 1500 years later. The forests of Europe became important centres of the charcoal-burning industry, and were gradually eaten away.

Coal was first dug up in Europe in the Bronze Age, about 2000 BC, but it was not until 1708 that Abram Darby, an English ironmaster, found a way of cooking coal to make coke. Outcrop coal was used first, but when it was all dug out miners began tunnelling into the ground to reach the seams. Colliers need light to work by, but their early lamps were of the open-flame kind, which often caused explosions by igniting the methane gas or 'fire-damp', prevalent in most mines. In 1816, the British scientist Sir Humphry Davy perfected a safety-lamp – an oil-lamp in which the flame is surrounded by a fine metal gauze screen.

Although air could enter the lamp to provide oxygen for the flame, the flame could not pass out to ignite the firedamp.

The first machine to cut coal was made by the British engineer James Anderton in 1868. It consisted of a revolving wheel-cutter driven at first by a steam engine, and later by compressed air. Today's coal cutters are driven by electricity and vary in type to suit the kind of coal and where it is found.

Coal can also be turned completely into gas. In the 1830s, *producer gas* was first made, by heating the coal in a limited supply of air. This process turned the carbon in coal into carbon monoxide, which could be burnt. Producer gas from coal has other gases – hydrocarbons – that make it richer, but it gives less heat than coal gas.

Nevertheless it is a useful gas in the steel industry.

A modern 'cat cracker', the business part of a big oil refinery. Catalytic cracking is a process which cracks, or splits, the molecules of petroleum in the presence of a catalyst — a substance which helps a chemical reaction but does not take part in it. The process was invented during the 1930s.

Above: A solar heating panel for domestic use, with, *above right*, a set of panels installed on the roof of a house.
Far right: Drilling for oil, 1970s style: the mobile platform 'Mr Cap' operating 240 km from shore in the North Sea.

Once started, the reaction to make producer gas keeps going from its own heat. In the Mond process, invented by the German-born chemist Ludwig Mond in the 1870s, steam is added to the air, and this makes a richer gas. If too much steam is added, the process needs extra heat from outside.

In the Lurgi process, developed in the 1930s in Germany, lump coal is heated with air and steam to make a gas for domestic use. Some tar and oils are also produced. At the same time and place, the Bergius process was invented. Powdered coal or tar is heated with hydrogen under pressure and produces a light oil suitable as fuel for motor-cars.

In the underground gasification process, the coal was not removed from the ground. Two holes were sunk to the coal seam, and a passage was burnt through the coal seam between them with high-pressure air or oxygen, or a current of electricity. In another method, shafts were sunk and galleries made, and holes were drilled through the coal. In both systems the coal seam was set on fire, and air, or a mixture of air and steam, was pumped through it and the gases collected.

Petroleum from natural seepages was used by the Sumerians from about 3000 BC. But most oil is found deep underground and, until the steam engine had been invented to drive power drills, this oil could not be exploited.

In 1859, Col. Edwin Drake drilled a hole 70 feet (21m) deep at Tinsville, Pennsylvania, using a steam engine to drive the drilling tool. This development really started the great petroleum industry and made modern technology possible.

Below: The miner's safety lamp, invented by Sir Humphry Davy in 1815-16. Davy refused to patent his device or to make money from it.
Below right: Operating modern coal-cutting machinery deep underground.

Steam

Hero of Alexandria, who flourished nearly 1800 years ago, was one of the great pioneers of steam. The model of the aeolipile, *below left*, has been constructed from contemporary descriptions of the device. The drawing, *left*, shows a steam organ, or hydraulis, also attributed to Hero, though the mechanism was known 400 years earlier.

The first known steam engine, the aeolipile, was made by the mathematician Hero of Alexandria about AD 100. It had a boiler to produce steam, which passed into a ball and escaped through bent pipes, making the ball rotate. Hero's invention has been described as the first jet engine, but it was in fact a reaction turbine. The idea was not taken up, and men had to wait until the late 1600s before steam was made to do work.

A practical steam engine was invented in 1698 by an English military engineer, Thomas Savery, to pump water out of mines. It had a boiler heated by a fire, from which steam passed to a cylinder. The cylinder was connected by pipes and valves to the mine-water. Some of this water was diverted to flow over the cylinder, causing the cylinder to cool and the steam in it to condense. As the steam condensed, it created a vacuum which sucked up the mine-water.

A Devon engineer, Thomas Newcomen, invented in 1705 an improved engine, in which the steam was cooled by a jet of water sprayed into it. The cylinder was mounted on top of the boiler, and had a piston on top. The piston was pushed up by steam pressure in the cylinder, and pulled down by the vacuum when the steam was cooled. The piston had a rod connected to one end of a beam which was pivoted in the middle. The other end was connected to a pump plunger.

Newcomen engines were very reliable, and some

Modern roads owe a great deal to steam. Steam road-rollers like this model of 1867 literally paved the way for the motor-car a few years later.

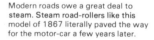

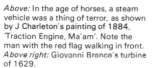

Above: In the age of horses, a steam vehicle was a thing of terror, as shown by J Charleton's painting of 1884, 'Traction Engine, Ma'am'. Note the man with the red flag walking in front.
Above right: Giovanni Branca's turbine of 1629.
Right: Steam on the rails: 'Locomotion No 1', which hauled the world's first passenger train on the Stockton and Darlington Railway in northern England in 1825.
Below: A model showing the operation of Thomas Newcomen's steam pumping engine.

were still used until quite recently to pump water out of coal mines. They were, however, very inefficient. It was left to a Scottish instrument maker, James Watt, to perfect the steam engine and turn it into a versatile source of power. Watt reasoned that a great deal of heat was wasted in warming the cylinder with the steam, and then cooling it again to condense the steam to water. So he devised a separate condenser to which the used steam passed from the cylinder, leaving the cylinder hot. He then linked the up-and-down motion of the piston in the cylinder to provide a revolving motion on a shaft which carried a flywheel. The flywheel kept the shaft turning between the strokes of the piston.

The crank, the most obvious way to turn to-and-fro motion into rotary motion, was the subject of a patent by a rival inventor. So Watt invented the sun-and-planet gear, which produces the same effect, going over to the crank when the other patent ran out.

Watt's engine meant that factories could be sited close to their raw materials and transport, instead of only near water, which had been used to drive the machines through waterwheels. It did more than anything to speed up the progress of the Industrial Revolution. But the Watt steam engine was still not particularly efficient, and various inventors turned to and improved it in the next hundred years. Jonathan Hornblower, another Briton, invented the compound engine in 1861. In this engine the steam is passed successively through two cylinders, operating at different pressures, thus doing two lots of work for the price of one.

The next breakthrough was the development of the steam turbine. Hero of Alexandria's primitive steam engine was a kind of turbine. The Italian engineer Giovanni Branca published a treatise in 1629 describing a machine in which a jet of steam played on the blades of a paddle-wheel, in much the same way as water falls on the paddles of a waterwheel. The first practical steam turbine was invented in 1831, by an American, William Avery. Avery's machines were efficient in their use of power, but they were noisy, hard to regulate, and liable to break down. The real importance of the steam turbine dates from 1884, when the English engineer Charles Parsons built the first multi-stage turbine — the equivalent, in turbine terms, of Jonathan Hornblower's compound piston engine.

Internal Combustion Engines

The simplest kind of internal combustion engine is the rocket, probably invented by the Chinese, whose armies used rockets, described as 'arrows of flying fire,' against the invading Mongol hordes of Genghis Khan in the 1200s. In a rocket, solid or liquid fuels are burned to make gases and the pressure of the gases drives the rocket along.

In another kind of internal-combustion engine, known as a heat engine, the heat of the gases is the main source of pressure, rather than the formation of the gases themselves. If the hot gas is passed out through a nozzle, you have a jet engine; if the gas moves a turbine's blades or a piston, you have a conventional internal combustion engine.

The first heat engine driving a piston was made by William Cecil of Cambridge, England, in 1821. He burned a mixture of hydrogen and air, producing water vapour, which he then cooled to produce a vacuum. This engine was slow and inefficient. Three years later the French physicist Nicolas Sadi Carnot published a theory of heat engines, stressing the need for high temperatures and pressures. But it was not until 1860 that his countryman Etienne Lenoir made a successful engine based on Carnot's ideas. Lenoir's engines were inefficient, but several hundred of them were sold.

Lenoir's engine was basically a two-stroke model — that is, one in which every other stroke of the piston is powered. Two years later, in Paris, Alphonse Beau de Rochas proposed a four-stroke cycle, in which every fourth stroke is powered. This is the system most generally used in car engines today, while the two-stroke cycle is mainly used for smaller engines, as in motor-cycles.

Beau de Rochas never built an engine to test his theory, but the German engineers Nikolaus Otto and Eugen Langen did in 1876. It was an immediate success — much more so than the two-stroke model the same pair had built ten years earlier. Both models were powered by gas, then the normal fuel for heat engines. But at this time large amounts of petroleum were being obtained from the ground, and it was found that the lighter oils obtained by refining, particularly petrol (gasoline), could be used to run gas engines, since they turned readily into combustible gases when mixed with air and heated by compression in a cylinder. The first engine using petrol as a fuel was invented by Siegfried Marcus in Austria in 1864. It was a two-stroke engine.

The real pioneer of the four-stroke petrol engine was the German engineer Gottlieb Daimler, who produced his first engine in 1885. Even more revolutionary were the engines of another German, Karl Benz. Benz's engines incorporated several novel features which are now standard in modern car design. The ignition was powered by an induction coil, which converted to very high voltage the current obtained from an accumulator. Benz also fitted his

Etienne Lenoir's gas engine of 1860. A slow and inefficient power unit, it was rated at a half horsepower.

The Wankel engine and its creator: *opposite,* Felix Wankel with an early example of the engine, and *right,* the four 'strokes' of a typical Wankel motor — 1 induction, 2 compression, 3 expansion, 4 exhaust. When the triangular piston has rotated once, it has completed the four-stroke cycle three times.

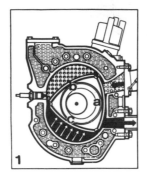 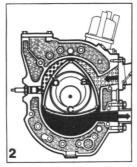 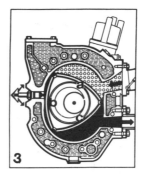 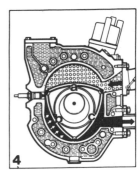

1 2 3 4

The ultimate in internal-combustion engines: the jet engine which Britain's Frank Whittle invented in 1938.

engines with removable sparking plugs. Another German, Wilhelm Maybach, invented the float-feed carburettor which keeps the fuel flow even.

If a gas is rapidly compressed, it becomes hot enough to ignite spontaneously. This self-ignition works better with heavier oils than petrol, which are also cheaper to produce. Several engineers began working on the idea of using these oils and the self-ignition principle. In 1873 an American engineer named Brayton invented a self-firing engine, started with the aid of a cylinder of compressed air. When running, the engine drove a compressor to recharge the cylinder.

But the most successful of all the heavy oil engines was produced by the German engineer Rudolf Diesel in the early 1890s. He also experimented with coaldust as a fuel, but abandoned it after a disastrous explosion.

The rotary piston engine was the brainchild of a German engineer, Felix Wankel, who invented it in the 1950s. In the Wankel engine, a triangular rotor turns eccentrically inside a combustion chamber shaped so that the three points of the triangle always form a seal with the walls of the chamber. The rotary piston is shaped with three dish-like depressions: as it rotates it forms, in effect, three pistons working in three cylinders.

The workings of a conventional four-stroke engine are shown in this exploded diagram of a Mercedes-Benz 2.8 litre 6-cylinder motor.

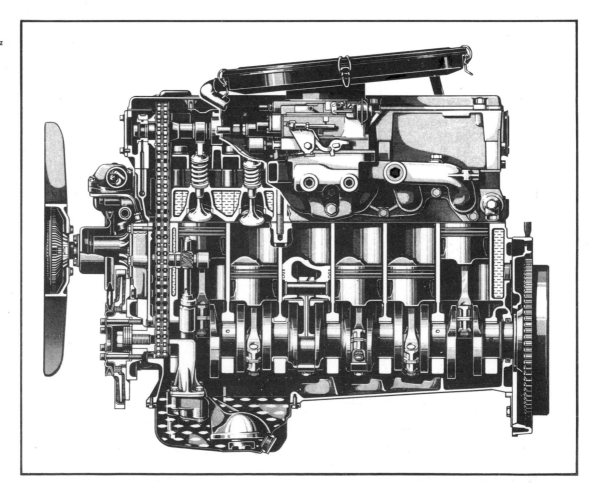

Electricity

The story of electric power begins back in the 500s BC, when the Greek philosopher Thales of Miletus found that amber rubbed with a piece of cloth attracted lightweight objects such as feathers. Around the same time a shepherd in Asia Minor (modern Turkey) discovered lodestones, when he found pieces of this magnetic substance sticking to the iron-shod end of his crook.

The first electrical device was invented by a German Otto von Guericke, in 1672. He charged a ball of sulphur with static electricity by holding his hand against it while rotating it on an axle. His experiment foreshadowed the theory, put forward in the 1740s by an English physician, William Watson, and the American statesman Benjamin Franklin, that electricity is in all matter and is transferred from one object to another by rubbing. Franklin believed lightning was a form of electricity, and to prove it flew a kite during a thunderstorm. He produced sparks from a key attached to the kite's string. As a result of thiis dangerous experiment, Franklin invented the lightning conductor, or lightning rod.

The first battery was invented by Alessandro Volta, an Italian aristocrat. He placed zinc and copper rods in salt water in a series of glass jars, and connected the zinc in each jar with the copper in the next. Touching

Above: An early form of electromagnetic engine, designed by the British scientist Sir Charles Wheatstone in 1840. Wheatstone based his work on earlier experiments.
Left: A power station of 1889, showing rope-driven alternators.

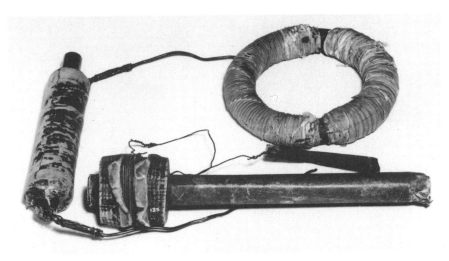

Above: Some of the original apparatus with which the British scientist Michael Faraday made his experiments on electromagnetism in 1831.
Right: Countries with fast-flowing rivers which can be dammed for hydro-electric power generate most of their electricity in this way. This dam is at Hinderfossen, in Norway.
Below: A reconstruction of Alessandro Volta's first experimental battery, the pile; Volta arranged a stack of silver and zinc discs, separated by cardboard saturated in brine; but in use the brine quickly dried out, so Volta switched to the more familiar arrangement of glass jars.
Below right: A series of huge stators — the static parts of generators — under assembly at an Oslo factory which manufactures power plant.

the two remaining rods, he received an electric shock. Volta's battery was a wet battery – one using a liquid. The ancestor of today's dry batteries was also a wet battery, the Leclanché cell, invented by the French engineer Georges Leclanché in 1865. It had zinc and carbon rods immersed in a solution of ammonium chloride. The modern dry battery contains a carbon rod in zinc chloride paste, surrounded by a zinc casing.

All dry batteries work on a chemical reaction. When the reaction is completed, electricity ceases to flow. In 1859 the French physicist Gaston Planché invented an accumulator, a lead-acid battery in which the reaction can be reversed by feeding power back, so that the battery can be used over and over again. The nickel-cadmium battery, which works on similar principles but is more durable, was invented by W Junger and K L Berg between 1893 and 1909.

No battery or storage cell can supply more than a small amount of power, and inventors soon realised that they needed a continuous source of current. The first electrical generator was devised by Michael Faraday, a Surrey blacksmith's son who became assistant to Sir Humphry Davy. In 1831 Faraday made a machine in which a copper disc rotated between the poles of a large magnet, with copper strips providing contacts with the rim of the disc and the axle on which it turned. Current flowed when the strips were connected.

The first working electric motor was made in the 1820s by William Sturgeon, of Warrington, Lancashire. He also made the first working electromagnet, and used battery-powered electromagnets in a generator in place of permanent magnets. Self-generating magnets which needed no battery to start them were produced by several inventors around 1866, including two English electricians, Cromwell Fleetwood Varley and Henry Wilde; Anyos Jedlik, a Hungarian physicist; and the pioneer American electrician, Moses Gerrish Farmer. But the first really successful model was the work of a German, Ernst Werner von Siemens. He produced his generator, which he called a dynamo, in 1867. Today, the term 'dynamo' is applied only to a generator which provides direct current. Simpler generators produce alternating current, and are called alternators. The development of motors which could work off alternating current was the work of an American engineer, Elihu Thomson.

Thompson also invented the transformer, which changes the voltage of an electric current and also its power. He demonstrated his invention in 1879, and five years later three Hungarians, Otto Bláthy, Max Deri and Karl Zipernowsky, produced the first commercially practical transformers.

Out-plated

Sometimes one invention can put paid to another. In 1712, Thomas Bolsover, a cutler of Sheffield, England, invented a way of fusing a thin layer of silver on to copper, making it possible to manufacture articles which looked like solid silver but cost a great deal less. The result is known as Sheffield plate.

But in 1832 another English cutler, George Richards Elkington, and his cousin Henry Elkington set out to see whether a film of silver or gold could be deposited by means of an electric current. Nine years later they established a large electro-plating works in Birmingham – and within a further nine years Sheffield plate was as dead as the dodo.

Smelting and Processing

The most outstanding inventors in the field of metal smelting and processing were probably those unknown men in the Neolithic Age who discovered that metal, when heated, could be shaped into tools, weapons and vessels. Their discovery was to bring the Neolithic Age to an end and lead eventually to the Iron Age. The methods of these first metalworkers were primitive. They used native metal – that is deposits of pure or relatively pure metal – for they knew nothing about extracting metals from ores. The metal was heated in small open hearths, probably fired by charcoal, and was beaten by stone hammers on stone anvils.

But as early as 4000 BC, Man's knowledge was so advanced that he had learned how to extract iron from its ore, and between 2200 and 1200 BC men developed short shaft furnaces with a blast created by bellows. The blast furnaces of today's great steel industries are highly sophisticated developments of the simple principles discovered by these ancient pioneers.

Extraction of the metal was achieved first in simple bowl furnaces. But it was soon discovered that if a wall were built round the bowl, creating a chimney, a fiercer draught and thus a greater heat could be obtained; and that if air tubes were thrust into the furnaces, even fiercer heat could be achieved. Archaeologists have discovered the remains of such short-shaft furnaces of the Sumerian, Assyrian and Chinese civilizations.

There is evidence that by 1400 BC the early ironmasters had learned most of the basic principles which govern the smelting and processing of iron today. The working of iron probably began in Asia Minor (modern Turkey) but soon spread, especially in Europe and Asia. By the Middle Ages, it was common in Spain and Central Europe and in England. The Spaniards developed the Catalan forge, in which the ore and the charcoal were heated in a large crucible and air was forced into the furnace by water-power. Another variety of furnace was the Stuckofen, which originated in Styria, in Austria, and had a shaft 10-14 feet (3-4m) high.

With the passage of time, many improvements were introduced. Water-power came into use for such processes as crushing, roasting and wire-drawing. In AD 1550 Hans Lobsinger of Nuremberg replaced leather bellows by wooden box-bellows. But it was the substitution of coal for charcoal that revolutionized the industry. It was a timely innovation. In England, for example, the demand for timber for shipbuilding and the voracious consumption of charcoal by the ironworks brought about an alarming deforestation.

In 1619 an English ironmaster named Dud Dudley claimed to have produced cast-iron in a coal-burning furnace; but it was to be another 100 years before another Englishman, Abram Darby, really perfected the process. In 1709, at the later famous ironworks of

Above: This contemporary wall painting of an Egyptian metalworker shows him using a blowpipe to make the furnace burn more brightly.
Left: A medieval woodcut showing a metalworker tending his furnace. Such primitive sources of heat were used by smiths and alchemists alike.

Below: Working drawings of the first Bessemer converter, which Henry Bessemer introduced at his steelworks in Sheffield in 1860.

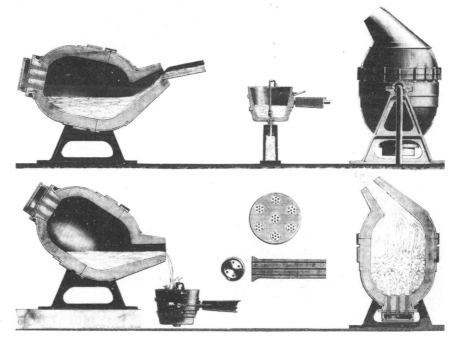

Far right: An iron foundry in 1822, showing little technical progress since ancient times.
Right: An ancient blast furnace discovered by archaeologists at Gyalar, in Transylvania. Similar primitive blast furnaces have been found in several parts of the Middle East.

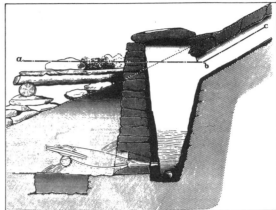

Coalbrookdale in Shropshire, he introduced new and successful techniques for smelting iron with coke and coal. Coke did not block the furnaces so much as charcoal, and it produced far greater heat.

A rapid consequence of Darby's invention was that iron-making moved away from the wooded areas of the Weald of Kent and the Forest of Dean in southern England, which were the sources of charcoal, to areas where coal was easily available. Thus the industrial areas of the Midlands and the North of England were created, and England became the scene of the Industrial Revolution.

Darby's revolutionary innovation was followed by the invention in 1740 by Benjamin Huntsman of a simpler and cheaper way of producing steel. Huntsman was originally a clockmaker from Doncaster, and his invention was, in effect, a re-discovery and adaptation of the old crucible process. It produced a better quality of steel. In 1783 a London ironmaster, Henry Cort, patented a method of expelling the carbon from molten iron by stirring it in a vibrating, reverbatory furnace – a method called puddling – which avoided contact between the metal and the fuel. And the following year Cort patented a rolling mill with grooved rollers, making the production of steel bars possible. There is some evidence that a similar process had been devised independently by Christian Polhem in Sweden 40 years earlier.

Many other improved methods of producing steel followed, particularly the invention by a Scotsman, James Beaumont Neilson, who realized that if hot instead of cold air were fed into the furnace, the blast and heat would be enormously increased. His discovery meant that three times the iron could be produced by burning the same amount of fuel — a considerable saving.

By the 1930s mechanization had greatly speeded up work in foundries, though some parts of the work were still carried out by hand.

The invention of the Bessemer converter in 1856 revolutionized the industry once more. Although the converter was eventually named after its developer, the British engineer Henry Bessemer, a similar method had been independently discovered in 1846 by an American, William Kelly of Kentucky, who made cauldrons for sugar-farmers. Kelly noticed that molten pig-iron in the bottom of the hearth was rendered much hotter by a blast of air when it was not covered by the charcoal in which it had been melted.

At about the same time that the Bessemer converter was patented, William Siemens, a naturalized Briton of German origin, and Emile and Pierre Martin of France, were working on the open-hearth furnace, the method now generally used in the American steel industry. Another significant development came in 1899 when Paul Héroult of France perfected the use of electricity in smelting. In his electric arc furnace, electrical resistance was passed through the bath of metal, a method similar to that used today.

Elusive metal

Although aluminium is one of the most common elements in the Earth's crust, it was not known as a pure metal until the early years of the 19th century. The breakthrough was made by the Danish scientist Hans Christian Øersted, who produced some pure samples of the metal in 1825. Twenty years passed before a German physicist, Friedrich Wöhler, invented a process for making the pure metal, and even then he could only produce pinhead-sized lumps of it. In those days knives, forks and spoons of aluminium were given as presents by royalty as something more precious than gold or silver.

The real breakthrough was the work of a French scientist, Henri Sainte-Claire Deville, who invented a process for extracting aluminium which for the first time enabled it to be produced on a commercial scale. Even then, it was still very costly.

The modern method, using electric current to break down the compounds of aluminium, was invented at the same time on opposite sides of the Atlantic Ocean – in the United States by Charles M. Hall, and in France by Paul Héroult. Their method precipitated pure aluminium from the compound alumina. That was in 1886. Two years later the German chemist Karl Bayer invented a process for producing alumina from bauxite ore, the main source of the metal – and the modern aluminium age was born.

Spinning and Weaving

Weaving is usually associated with the manufacture of cloth or textiles, but its origin is older and more basic than that of spinning thread, the preliminary to weaving cloth. Weaving is in fact the method of interlacing to form a solid material, and archaeological discoveries suggest that Man first invented weaving to make baskets and mats from vegetable fibres.

The perishable nature of the basic materials used in weaving, whether in basketry or the manufacture of cloth, means that very early records are few. The earliest evidence of basketry dates from about 5000 BC in Egypt and Iraq, but the remains of a fishing net made 2000 years earlier still of twined bast threads have been found. The principles of basketry and weaving are basically the same, with the material being interwoven in opposite directions. But while the techniques employed in basketry have changed little, Man's inventiveness has revolutionized the manufacture of textiles.

The first invention was that of the spindle, which enabled Early Man to draw and twist flax, cotton and wool to make an elastic and strong thread. The first spindle consisted of a shaft of wood or bone passed through a disc of stone or clay, making the spindle look something like a medieval dagger. In the top of

the 'hilt' was a groove through which the thread passed.

Looms were known in ancient Egypt, and at first they lay flat along the ground. Upright looms, in which the warp threads are vertical, were developed soon after, and proved more convenient to work.

It was the 1700s that saw revolutionary changes in the manufacture of fabrics. The first change was the

Top left and right: Spinning and weaving as they were carried on in ancient Egypt. The Egyptians produced a variety of textiles, some of them gossamer-fine.
Above: Samuel Crompton's spinning mule, which combined features of James Hargreaves' spinning jenny and Richard Arkwright's spinning frame.

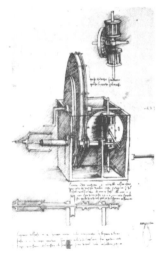

Above: That most prolific of all Renaissance inventors, Leonardo da Vinci, left drawings for a machine for spinning yarn more than 200 years before anyone else produced one.
Left: When the loom dominated the home: this interior of a silk weaver's cottage at Lyon in France was typical of many in all branches of textile weaving.

John Kay's flying shuttles, of which several kinds are shown here, completely revolutionized the textile industry.

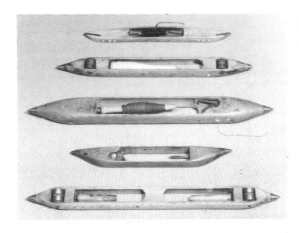

Below, top: Operating a modern replica of James Hargreave's spinning jenny.
Below, bottom: A vertical warping mill, with which the longitudinal threads of a piece of fabric were prepared for the loom.
Below right: A reproduction of fulling stocks, a device which was employed to shrink woollen fabrics, making them thicker and more compact.

flying shuttle, invented by an Englishman, John Kay; it speeded up the process of weaving enormously. In about 1764 another Englishman, James Hargraves, invented the 'spinning jenny' so that the speed of spinning thread could keep up with the rate of weaving. 'Jenny' was a corruption of the word 'engine'.

Five years later another Englishman, Richard Arkwright, invented a spinning frame, which had the advantage of making a strong thread from cotton which could be used for warp threads; up to then, warp threads were always made from linen because no cotton thread was tough enough. Arkwright's machine was driven at first by horsepower, and later by water-wheels. An improved spinning machine was invented in 1779 by Samuel Crompton, a combination of the best features of Hargraves' and Arkwright's machines whose hybrid origin gave it the name 'mule'. Four years later Eli Whitney, an American, devised a cotton-cleaning machine which led to his country's becoming the largest cotton producer in the world.

At the end of the 1700s English manufacturers were looking for a simple loom which would weave in patterns, and offered a prize for its invention. The prize was won by a French weaver's son, Joseph Marie Jacquard of Lyon, who perfected between 1801 and 1804 an automatic loom which would weave intricate patterns. It was operated by paper cards with holes punched in them, very much like the punched cards of modern business machines. The essential principles of the Jacquard loom are still used in the manufacture of textiles today.

The greatest advance of recent years has been the invention of man-made fibres, produced largely by chemical means. The first successful man-made fibre was invented in the 1880s by a Frenchman, Count Hilaire de Chardonnet. Its main ingredient was cellulose, the basic material of wood. Chardonnet called his invention 'artificial silk', but it is better known today under its later name, rayon. Another wood cellulose fibre, Celanese, was invented after World War One by two Swiss brothers, Henry and Camille Dreyfus. They devised it to find a use for large quantities of cellulose acetate which was left on their hands when the war ended. The acetate had been used for varnishing the fabric of aircraft. Nylon, the first of the modern fibres, was devised in 1935 by chemists of the E.I. du Pont de Nemours Company in the United States.

Pacemakers

Before the Industrial Revolution, spinning and weaving were cottage industries. Spinning was done by the women-folk and children of a household, while the man of the house tended his smallholding. In the evening he would go to his loom and weave in an hour or two all the thread his family had produced during the day.

When John Kay made his flying shuttle in 1733, he upset the routine — the spinners just could not keep up with the weavers. But after the development of Hargrave's spinning jenny and Samuel Crompton's mule, the weavers could not keep pace with the spinners. Not until Edmund Cartwright produced his power looms in 1786 was the industry back in balance — and by that time, invention had taken it out of the cottages and into the factories.

OI–H°

Construction

This bas-relief from the tomb of Hammurabi the Great, king of Babylon, who died about 1913 BC, shows a primitive crane in operation. Note the slaves in the treadmill which powered it.

Man's first homes were trees and caves. But as time went on he needed more shelter than these could provide, and so he began to build simple huts of reeds covered with woven plant fibres or the skins of animals. The next stage was to daub the reed framework with mud and clay, which hardened when dry. Large-scale construction using stone and timber, dates from about 6000 BC. It began in the Middle East, and spread to ancient Greece and Crete.

An important invention in this type of building was the post and lintel – two upright stones with a third crosswise above them. Stonehenge, on Britain's Salisbury Plain, is an example of this kind of structure, built before 2000 BC. But the post and lintel is limited in span by the size and strength of the stones available. The arch, which can span larger distances, was probably invented in the ancient city of Ur in Mesopotamia around 1400 BC.

The earliest arch found at Ur spanned a gap of only 32 inches (80 cm), but it had another novel feature: it was built of bricks. The first bricks were made of dried clay about 8000 years ago, in the same part of the world. The first kiln-fired bricks were made about 1400 BC. Roofing tiles were probably the invention of the Greeks around 600 BC.

Cement, made from limestone and gypsum, was another Middle East invention. The Romans developed the material a stage further by inventing concrete, which they made from possolana, a sandy volcanic soil. They used it as a basic structural material for domes, vaults, and walls, which were faced with brick or marble. The Romans also used glass – invented by the Egyptians about 1500 BC – for windows.

Metal was not used extensively in building before the Industrial Revolution, except for bronze, copper and lead, used for waterproofing roofs. The first building in which iron formed part of the fabric was the church of St Anne, in Liverpool, England, built in 1772. Cast-iron pillars supported the structure. Iron was also used to build bridges, the first of which was

Opposite: Reinforced concrete has many applications in the construction industry today. Under construction here are the legs of the 155m-tall Barton Pier of the new Humber Suspension Bridge, the world's longest.

The world's first cast-iron bridge at Coalbrookdale, Shropshire, erected by John Wilkinson and Abram Darby in 1779. It was in regular use until the 1950s, and is now a national monument.

Above: A water-powered tool, Hydrotool, developed in the United States in the 1970s as a spin-off from space technology. It has many applications in the construction industry.

erected in England at Coalbrookdale, Shropshire, in 1779, and is still standing.

In the mid-1800s a Scot, Robert Mushet, added manganese to steel and found it made the steel less brittle. As a result, the metal could be rolled and shaped into girders. The invention of steel girders opened the way for the skyscraper. Previously, any building taller than about five storeys required extremely thick walls to carry the load. In modern buildings, the steel skeleton carries the load, and the walls are simply added like curtains. The first skyscraper was built in Chicago in 1883.

In 1824 an English builder, Joseph Aspdin, invented portland cement, a mixture of limestone and clay. A French lawyer, Jean-Louis Lambot, discovered that concrete could be greatly strengthened by casting it around iron rods, and in 1867 another Frenchman,

Joseph Monier, patented a version of Lambot's invention using portland cement. This reinforced concrete, with steel girders, made really tall buildings possible, and the first skyscraper using this method of construction was built in Chicago in 1903.

Pre-stressed concrete, in which the steel reinforcement is stretched while the cement around it sets, was devised by Lambot in 1849 and came into general use in the 1900s. It has made possible such exciting buildings as the Sydney Opera House and Rome's Palace of Sport.

Many of the tools used in construction are so old no-one knows when they were invented. Cranes were in use in ancient Babylon more than 3700 years ago, but they remained simple man-powered machines until the early 1800's, when steam power was applied to them. Even the pneumatic drill was invented – in principle – as long ago as the 2nd century BC, by the Greek physicist Ctesibius; but it was not brought into use until the 1860s.

Pharaoh power

To build a structure like the Great Pyramid today would take hundreds of men using elaborate cranes, railways to transport the blocks of stone, scaffolding, cement-mixers – and, above all, power from electric motors and internal combustion engines.

When the Great Pyramid was built at Giza, in Egypt, more than 4000 years ago, none of those modern aids was available. Yet the Pyramid *was* built — with 2,300,000 blocks of stone, each weighing about 2.5 tonnes.

How was it done? The Pharaoh Khufu, who ordered the Great Pyramid to be built, had two things to take the place of modern inventions, enormous resources of manpower and plenty of time.

Above: Construction means noise — but the quiet pile-driver, developed by British engineers in the 1960s, goes a long way to making life more pleasant.
Right: New methods, new shapes — concrete again made possible the exciting roof of the Sydney Opera House in Australia.

Chemicals

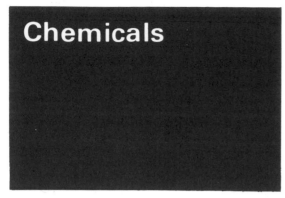

The popular picture of an inventor is a man of metals, of wheels and springs and levers and machines that go 'whirr' and 'click'. But though such inventors did exist, their work forms only a small part of the total history of invention. And an invention is not necessarily a gadget: it may well be a way of doing something or making something.

Probably the most magical of all inventions are those involved in the chemical industry, and in its early days chemical invention did contain a good deal of magic. For early chemists were mostly alchemists, whose ambition was to transmute base metals, particularly lead or copper, into gold and silver. The first true alchemists were in action about 1900 years ago.

Medieval alchemists used a very wide range of chemical substances, and some which a modern chemist would probably avoid; Geoffrey Chaucer in 'The Canon's Yeoman's Tale' describes some typical materials:

> Arsenik, sal armoniak, and brimstoon,
> And herbes coulde I telle eek many oon . . .
> Unslekked lym, chalk, and gleyre of an ey,
> Poudres diverse, asshes, dong, pisse, and cley,
> Cered pokets, sal peter, vitriole . . .

'Gleyre of an ey' is white of egg, and 'cered pokets' are small waxed bags. Many other substances were also used, and Damian the Florentine, court alchemist to James IV of Scotland, favoured large quantities of 'aqua vitae', which was probably whisky – and no doubt stimulated his inventiveness greatly!

The modern chemical industry sprang from, and contributed to, the Industrial Revolution. All industries need chemicals for processing other substances, and indeed the quantity of sulphuric acid an industry uses is still a measure of the size of its industry. It is essential for producing soda, making dyes, fertilizers and paints, processing leather, and refining oil and many metals.

The lead-chamber process for making sulphuric acid was invented by a Birmingham, England, physician, John Roebuck about 1845. The process took place in large lead-lined tanks, and remained in use until 1914. It was replaced by an even better process invented by a Bristol vinegar manufacturer, Peregrine Philips.

Soda is as important as sulphuric acid for industry, and the great process for making it was invented in 1791 by a French surgeon, Nicolas Leblanc. He won a 100,000-franc prize for it, though he never saw the money because the French Revolution overtook him before he could collect it. The Leblanc process was also replaced by a better one, the Solvay process, invented by a Belgian chemist, Ernest Solvay, in 1863.

In the modern world, probably the most familiar of all substances are plastics, which are entirely the work of chemists. One of the pioneers of artificial plastics was the German chemist Christian Schönbein, who tried to make articles from nitro-cellulose in 1845. The substance proved highly dangerous, and became better known as a high explosive under the name of guncotton.

In 1865 a British metallurgist, Alexander Parkes, made a plastic substance from nitro-cellulose, camphor and alcohol, which he called Parkesite, later modified by his associate Daniel Spill and named Xylonite. An improved version of this substance was Celluloid, invented in 1869 by John W. Hyatt, a printer of Albany in New York State.

The next step forward was taken by a Belgian-born American chemist, Leo Hendrik Baekeland who, in 1909, combined carbolic acid and formaldehyde to make Bakelite, the first true synthetic resin. It was prepared in powder form in 1916 so that it could be moulded. In the 1920s and 1930s chemists prepared a number of other substances which had the advantage over Bakelite that they could be made in clear bright colours, while Bakelite was only available in dark colours. It was at this point that the whole plastics industry really began.

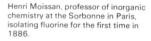

The mass spectrometer, an electronic device which separates atoms, molecules, and molecular fragments by their masses, is an important tool of the modern chemist, who uses it to analyse substances. The mass spectrometer is the result of a series of inventions beginning in 1898 and still going on.

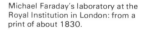

Henri Moissan, professor of inorganic chemistry at the Sorbonne in Paris, isolating fluorine for the first time in 1886.

Michael Faraday's laboratory at the Royal Institution in London: from a print of about 1830.

Far right: Just some of the circuit boards in a computer of the 1970s.

Electronics

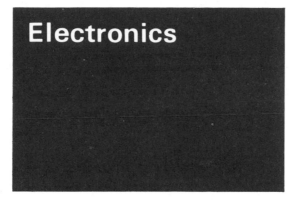

Above: A laser interferometer, an electronic device which uses laser beams for measuring to extremely fine limits of accuracy, and also for calibration of industrial machining equipment.

The story of electronics probably begins with a German glassblower, Heinrich Geissler. Around 1860, in the course of some experiments Geissler removed most of the air from a glass tube and passed an electric current through it. The tube glowed with a brilliant fluorescent light. Geissler had, without realizing it, invented the ancestor of the neon light, and also of the television tube.

The next invention was by the British physicist William Crookes, in the late 1870s. Crookes found that an obstacle placed inside a Geissler tube would cast a shadow on the other end. Crookes also discovered that a magnetic field would affect the behaviour of the electric current inside a Crookes tube, as his device is still called.

The cathode-ray tube, the heart of all television sets, was perfected by the German physicist Karl Ferdinand Braun in 1895. It consists of a funnel-shaped glass bulb from which most of the air has been pumped out, like the Geissler and Crookes tubes. In the neck of the tube are two electrodes (electric terminals), the cathode and the anode. The disc-shaped cathode is connected to the negative side of a high-voltage supply, and the cylindrical anode to the positive side.

Radiation, called cathode rays, flows from the cathode and strikes a screen on the face of the tube, producing a greenish-yellow flourescence.

Also in 1895 Wilhelm Roentgen, a German physicist, discovered X-rays, using a Crookes tube, and the British physicist J J Thomson invented a device for deflecting the rays inside a cathode ray tube.

The next stage was the invention of an electron tube, called a valve, by a British electrical engineer, John Ambrose Fleming, in 1904. His valve conducted current in one direction only – hence its name. This property enabled it to be used as a rectifier, for changing alternating current electricity to direct current, and for detecting incoming radio signals. Further improvements to valves were invented by the American Lee de Forest in 1906, the German Walter Schottky in 1916, and the Dutchman Benjamin Tellegen, in 1926.

Valves remained the mainstay of radio and TV until the development of transistors. Transistors are made of materials called semiconductors, because they conduct electricity but not so well as metals such as copper or iron. The first transistor was invented by three scientists working for the Bell Telephone Company in the United States in 1947 – John Bardeen, Walter Brattain and William Shockley. Unlike a valve, a transistor does not have to be heated, and uses very little electrical power, and it can be made very much smaller than a valve. These two great advantages led to the development of small transistor radios, and reduced the size of complicated electronic devices such as computers from several rooms to one large cabinet.

The latest developments with transistors are called integrated circuits. Several transistors with all their associated wiring and other devices, can be formed in a single 'chip' of semiconductor material only 1 mm square. As a result, engineers can make really miniature circuits. Chips of this kind are the basis of such devices as pocket calculators. Semi conductors are also sometimes used in another electronic gadget, the maser – an abbreviation of 'Microwave Amplification by Stimulated Emission of Radiation' – which is used to amplify extremely short radio waves. The maser was invented in 1953, and the laser, which amplifies light waves to produce intense beams of light, five years later. Charles Hard Townes of the United States and two Russians, Nikolai Basov and Aleksandr Prokhorov, shared a Nobel prize for this development.

Other electronic devices are based on materials which generate an electric current when light shines on them. They include photoelectric cells which open doors or trigger burglar alarms, and the TV camera tube, which was invented by the Russian-born American physicist Vladimir Zworykin in the 1930s.

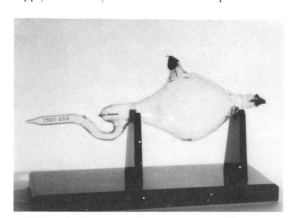

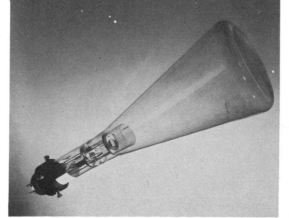

Above: A vast panel of solar cells which provide the power for an Intelsat communications satellite.
Right above: A Crookes tube of 1879, a device which led to the discovery of X-rays.
Right: An early cathode-ray tube, the 'heart' of all TV sets and oscilloscopes.

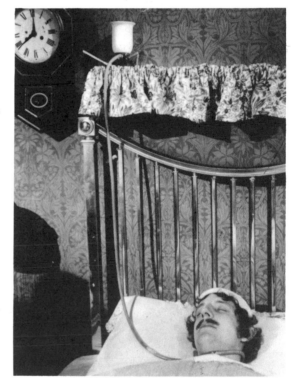

Inventors have done a great deal to make the world a healthier, faster, more efficient and even more dangerous place. But like any other group of people, they are sometimes good for a laugh.

Fortunately for us, most of these inventors have had their work immortalized in the outwardly sombre files of the world's patent offices. The word 'patent' is short for 'letters patent', an old English term for a document or documents granting a special privilege or monoploy. The patent system began in England, and the oldest patent known was issued in 1449 by King Henry VI. It was given to a Fleming, John of Utynam, for the process by which he made the stained glass windows for Eton College.

However, the modern patent system owes its origin to King James I of England, who tidied up the laws regarding monopolies in 1624. Up to that time, English sovereigns were in the habit of giving monopolies to various court favourites, with the result that there were many abuses. By the law of 1624, James banned all monopolies, except for inventors of new manufactures, who were to be allowed the sole rights to their inventions for 14 years. The term is different now – 16 years in Britain, 17 in the United States and Canada, and up to 20 in some other countries – but the principle is the same.

The inventor has to keep on paying fees for every year the patent is in force. So, unless he is very well off, or has acquired a ready market for his invention, he is likely to let it lapse. It is unlikely, for example, that the British inventor who patented a scheme for having the cremated remains of your nearest and dearest pulverized and turned into umbrella handles found a ready market for his idea.

The world's weird inventions fall into three main groups: impractical or just plain crazy ways of attaining a sensible object; sensible methods to attain a crazy object; or just plain crazy all the way. In the past, people have sometimes had difficulty in deciding if an idea was crazy or not; for example, Sir George Cayley's coachman, having flown in his master's newly-invented glider, would probably have come down heavily on the side of calling the idea absurd. On the other hand, otherwise shrewd business men have been persuaded to invest money in many of the machines for perpetual motion, which cannot possibly work.

Although the physical principles of motion have been known and understood for years, inventors keep

on producing machines for perpetual motion, and patent offices keep on turning them down. The idea behind a perpetual motion machine is simple enough: in theory the machine, once started, will run continuously without any change or any external power source. Inventors have toyed with counter-balances, weighted wheels, and all sorts of devices, but they always come against one insoluble problem: friction. Any machine working on Earth is bound to have some friction trying to stop it, even if it is only the resistance of the air around it. Since friction absorbs energy, any machine must eventually stop unless more energy is fed into it.

The nearest Man has come to achieving perpetual motion is to put a space probe into orbit around the Sun. With no air to slow it down, the movement of the space probe can continue apparently indefinitely – at least, until the Sun dies.

Not surprisingly, transport has produced its own crop of weird ideas. Among balloons perhaps the most original was that designed by a Parisian, Charles Wulff. In 1887 he produced a design for a balloon in which the motive power was to be supplied by large birds such as eagles, tethered to the framework of the machine. Other designs for balloons required manpower for propulsion, at least overcoming the problems that have always faced Man when he has tried to build wings and fly. For Man is much heavier-boned in proportion to his size than a bird, and therefore needs more power to lift him into the air than his arms or legs can provide.

Among the earliest pioneers of heavier-than-air machines was a Brazilian priest and mathematician, Batholomeu Lorenco de Gusmao. He built a model, called the Passarola, which was a cross between a bird and a boat. In 1709, de Gusmao tested his parachute-sailed craft in Lisbon, and to most people's surprise it actually flew. The priest was not, however, sufficiently convinced by the test-flight to put his faith in a man-sized version.

Another strange aeroplane which never got off the ground was the brain-child of a Glasgow engineer, Joseph Kaufmann. It was a steam engine with wings that flapped like those of a bird. A long rod with a heavy weight at the end of it was suspended beneath the machine. The purpose of the weight was to keep the craft on an even keel. Kaufmann reckoned his machine could travel through the air at more than 50 mph (80 kph), carrying its passengers in three gondolas. He built a small model of the machine in 1869, but when it was tried out on the ground, the wings flapped so violently that the whole thing fell to pieces.

Wheeled vehicles have also attracted their share of strange inventions. Perhaps one of the oddest of these was patented in 1868 by two Newark, New Jersey, inventors, Zadoc P. Dederick and Isaac Grass. It was only partly wheeled, because it was built in the form of a mechanical man pulling a cart. The mechanical man was powered by a steam engine concealed in his body, and was intended to walk forward with strides of his jointed legs.

The early days of bicycles and tricycles also produced some oddities. One novel machine was the 'Sociable' of 1882, on which the riders sat side by side between two enormous wheels, steering with a smaller wheel out in front. About 14 years later another resourceful inventor produced a Companion Safety Bicycle, inspired by the Sociable. This machine had the modern 26-inch (66 cm) wheels, and the riders sat side by side, each gripping a pair of handlebars which steered the machine through a system of levers. Maybe it was a difference of opinion between two pioneer riders which inspired a later version having one set of handlebars only. The non-steering rider carried a sporting gun, presumably to bag some game as the pair cycled along!

The first true tandem machine was built by a British inventor, M D Rucker, in 1884. It consisted of two 59-inch (1.5 m) wheels from 'penny-farthing' bicycles linked together, and required an amazing degree of acrobatic skill to manage. Even so, it was popular for several years in the United States and Germany.

The motor-car has also been subject to some original ideas. Some of the most ingenious notions were based on the idea of the treadmill or the dog-operated spit. A carriage devised in 1870 by F H C Mey, of Buffalo, New York, had a pair of dogs running inside the front wheel to propel it. A more sophisticated version of the same basic principle was the Dynasphere, built in the 1930s by an English engineer, Dr J A Purves. This monowheeled vehicle rolled along like a ball. The driving seat and engine ran on rails inside the giant wheel, like the dogs in the treadmill. To corner, the driver moved the rails to one side or the other, tilting the wheel; the driving unit stayed upright. Braking was achieved by switching off the engine. The Dynasphere could travel at almost 30 mph (50 kph), and ran as easily backwards as it did forwards.

Several inventors produced a vehicle in which a horse trotted along an endless belt inside the machine.

Inventors have in their time thought of everything, including walking on the water! This machine was devised by Johann Cristoph Wagenseil, described as a German of 'considerable rank and erudition', in the 1800s. It was called the hydraspis, or water-shield, and was said to be so simple that any carpenter or smith could make it.

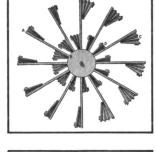

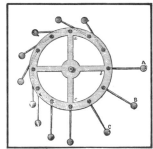

Above left: This invention of the 1970s was called a 'muscle powered personal transportation vehicle' or Supertrike for short. It was obviously designed for those who like to sit back when taking exercise.
Above: Two examples of perpetual motion machines; that at the top was devised by Bernard Launy, and the one in the lower picture by Jeremie Metz. Like all such machines, they were doomed to failure.

Far left: If a drawing can be called an invention, then a French cartoonist of 1818 invented the motor-cycle! This fanciful machine was intended as a caricature of the Drasine, and the artist gave it the jaw-breaking name of 'Vélocipédraisiavaporianna'.
Left: This combined gun and plough was invented in the United States in 1862 for use on the Western Frontier, where men wuz men and liable to turn mean and ornery any time.

Lighthouses were highly developed during Victorian times — but few so highly original as this floating version. Despite its four anchors, it seems unlikely to have been a very comfortable place.

water pump or any other useful machine. A weight-operated scraper kept the moving platform clean, and the manure thus won was dropped straight into a cart to enrich the fields.

But it is probably in the realm of home and household gadgets that the inventor has produced the most eccentric ideas. One anonymous invention of 1873 was designed to harness 'hitherto wasted female power'. A weighted arm suspended from the ceiling was connected by a system of cords and pulleys to a rocking chair, in which the wasted female sat. As she darned her husband's socks, she rocked to and fro causing the counterweighted arm to churn the butter and rock the baby's cradle. The inventor called his device 'The New Domestic Motor'.

A more practical device was invented about 1884 by a French engineer named Daussin. He made a miniature steam-engine which could be placed on the kitchen range. Its boiler held about a litre of water, which rapidly heated up and provided steam to turn a flywheel. A belt drive from the flywheel could be used to power a sewing machine.

Two other ingenious ideas for saving labour came from North America. Sinclair Arcus of Chicago made a combined washing machine and rocking chair. As with the New Domestic Motor, rocking powered the machine. Remove the chair and the machinery, and you had a useful tub for the weekly bath. Cover the whole apparatus with a flat top, and there was the dining table. Another device, produced two years earlier in 1888, used a child's swing to operate a butter churn.

The comfort of the ladies has not escaped attention. One English inventor recently put forward an idea for holding up women's stockings with a pair of helium-filled balloons, attached to the outside of the stockings. What the wearer did about the unsightly bulges produced by the balloons was not included in the inventor's specification.

Bed is also a frequent subject for inventive talents.

Many inventors were worried that ordinary flanged wheels would not provide sufficient grip on those new-fangled railways — hence such fantasies as 'Prosser's Patent Guide Wheel Carriages'.

One of these was a streetcar, invented by Augustus W Getchell of Cleveland, Ohio, in 1887. The same idea was applied by an English inventor of a few years earlier to keeping horses fed and clean in the stable. His invention called for a moving platform fitted inside the horse's stall in place of the floor. The faithful animal could be set at a steady trot, and an ingenious arrangement of pulleys, belts and levers led the horsepower to drive a straw-baler, a flour-mill, a

For the little woman whose husband snores, a British inventor in the 1870s proposed a harness of padded straps to keep the sleeper's mouth shut. He appears to have based it on the scold's bridle, used on talkative women in medieval times.

In 1872, long before the days of electric fans, J B Williamson of Louisville, Kentucky, made a clockwork bedroom fan which could be driven either by a spring or by weights. An arm rocked to and fro above the bed, carrying on either end a series of ribbon-like whisks to keep the air moving – and, incidentally to scare away mosquitos.

However, comfort in bed is one thing: getting up in the morning is quite another, and that problem has exercised inventors frequently. Devices for arousing people from their slumbers range from the bed that tips up and deposits its occupant straight into a healthful cold bath to devices that hit him over the head or shower him with corks. More subtle are a whole range of gadgets for trickling drops of cold water down his neck, all operated by the invaluable alarm clock.

If you think that the alarm clock itself is a sufficiently barbarous device, take comfort from the fact that the first such clocks were apparently devised in the Middle Ages to arouse monks for services in the middle of the night.

Warfare also produces its crop if zany ideas – only sometimes the crazier the idea, the more use it has in real life. Back in 1916, Jones Wister of Philadelphia, Pennsylvania, invented a curved attachment which could be fitted to the end of a rifle or a machine-gun, so that the user could fire round corners. People laughed at the time. But during World War Two, German infantrymen engaged in street fighting in Russia were armed with the 'Krummlauf', a machine-pistol having a curved barrel for firing around the

corners of buildings. Since then, American and Russian versions have been made and used in action.

During World War Two, many devices were produced for special purposes, such as the floating harbours used in the Normandy Invasion of 1944, amphibious tanks – then a novelty – and midget submarines for underwater sabotage. But many other ideas were the subject of experiment, by hard-headed men with a war to win, and some of them were as weird as any you will find in the patent offices.

Lieutenant-Commander Nevil S Norway, RN – better known as Nevil Shute, the novelist – spent many uncomfortable months working on a device called the Great Panjandrum. The brainchild of a Group-Captain Finch-Noyes, the Great Panjandrum looked like a giant cable-drum, with iron wheels 10 feet (3 m) across. The centre of the drum was to be filled with TNT. The Great Panjandrum was designed to roll forward over an enemy beach until it reached the defences of the Atlantic Wall, a huge concrete structure which it was believed the Germans had built all along the coast of France (in fact, it did not exist). Then the TNT would be set off, blasting a way through for the invading troops.

The device was to be propelled by a series of rockets set around the rims of its huge wheels, like a pair of giant Catherine wheels. It was steered by cables unwinding from the centre axle, where the TNT was stored. Extensive trials were made on beaches in the West of England, but the Great Panjandrum proved to be dangerous to everyone and everything except a potential enemy. Not only was it difficult to steer, but also the rockets broke loose and hurtled across the beach to the peril of all onlookers.

Even more bizarre was an idea, put up in all seriousness by a British civil servant, for combating flying bombs. The inventor's plan was to send up bubble-making apparatus attached to barrage balloons. He hoped that the flying bombs would suck the bubbles into their pulse-jet engines and explode. More serious was the scheme to camouflage the River Thames – which gleamed brightly in the moonlight and provided a guide to German bombers – by spraying it with coaldust. Alas, tides in the river estuary broke up the layer of coaldust as fast as it was laid. However, the scheme's inventors had the satisfaction of knowing they were on the right lines. They camouflaged a canal, where there was no current or tide, so successfully that an elderly man exercising his dog walked into the water under the impression that it was a tarmac-covered road.

Far left: The early days of motor-cars (or should it be motor-cycles?) produced a crop of very strange vehicles, such as this Riley tricar of 1904. The passenger appears to be used as a form of battering ram. *Above:* A cycle railway is possibly one answer to high costs of fuel and wages. One was apparently constructed in the United States in the late 1800s, and was said to provide a much smoother and pleasanter journey than travelling by road.

Far left: Bowling along, 1930s style — the Dynasphere car invented by an English engineer, Dr J A Purves. *Left:* The Otto dicycle was adopted — and adapted — for a time by the British Post Office during the mid-1800s.

Index

Index

SIGNIFICANT FIRSTS

Acknowledgments

We are indebted to the following organisations and individuals for permission
to reproduce photographs and other illustrative material:

Albright and Wilson Limited
Associated Newspapers Limited
Audi/NSU
Australian News and Information Bureau
Bang and Olufsen
British Aircraft Corporation
British Broadcasting Corporation
British Hovercraft Corporation
British Rail
Canadian Pacific Railways
Central Electricity Generating Board
Central Office of Information
Coal Board
Danish Government
D H B Construction Limited
Egyptian State Tourist Administration
Esso Petroleum Limited
First Features Limited
Ford Motor Company
Fox-Rank Distributors
French Government Tourist Office
General Dynamics Corporation
Gillette Industries Limited
Greater London Council
Grundig (GB) Limited
Gutenberg Museum
Hastings Public Library
Hewlett-Packard Limited

H M Stationery Office
Hoover Limited
Hovercraft Development Limited
John Howard and Company Limited
IBM
Imperial War Museum
International Computers Limited
Irvin GB Limited
Japan National Tourist Organization
Lloyds Bank
Lucas Industries Limited
Mansell Collection
Marconi Company Limited
Martin Baker Aircraft Company Limited
Masini-Franklin Partnership
Mercedes-Benz
Miehle-Goss-Dexter Incorporated
Ministry of Defence
National Farmers Union
Nigerian High Commission
National Aeronautics and Space Administration
National Physical Laboratory
Norwegian Government
National Trust
Novosti Press Agency
Parker Pen Company
Philips, Eindhoven
Post Office

Pultron Industries Incorporated
John E Ray Collection
Rolls-Royce
Theodore Rowland-Entwistle
Royal National Institute for the Blind
Royal Navy
Royal Radar Establishment
Sandrock Marine
W A Sheaffer Pen Company Limited
Schick Incorporated
Science Museum, London
Shell International
Singer Company (UK) Limited
Southern Television Limited
Southern Water Authority
Spanish National Tourist Office
Taylor Woodrow Construction Limited
Thorn Domestic Appliances (Electrical) Limited
H M Tower of London Armouries
Tracked Hovercraft Limited
United States Information Service
United States Navy
Vickers Oceanics Limited
Volkswagen (GB) Limited
Westland Aircraft Limited
Westrex Company Limited
West Yorkshire Folk Museum
Wilkinson Sword Limited